直-9直升机航电维护图解
（中英对照版）

A Chinese-English Visual Manual for Avionic Maintenance of Z-9 Helicopters

主 编 ◎ 吴立勋 徐 希 程新平

北京理工大学出版社
BEIJING INSTITUTE OF TECHNOLOGY PRESS

图书在版编目（CIP）数据

　　直 - 9 直升机航电维护图解：中英对照版 / 吴立勋，
徐希，程新平主编. —北京：北京理工大学出版社，
2022. 4

　　ISBN 978 - 7 - 5763 - 1233 - 1

　　Ⅰ. ①直…　　Ⅱ. ①吴… ②徐… ③程…　　Ⅲ. ①直升机
- 航空电气设备 - 维修 - 图解　　Ⅳ. ①V275 - 64

中国版本图书馆 CIP 数据核字（2022）第 066188 号

出版发行 / 北京理工大学出版社有限责任公司
社　　　址 / 北京市海淀区中关村南大街 5 号
邮　　　编 / 100081
电　　　话 / （010）68914775（总编室）
　　　　　　（010）82562903（教材售后服务热线）
　　　　　　（010）68944723（其他图书服务热线）
网　　　址 / http：//www. bitpress. com. cn
经　　　销 / 全国各地新华书店
印　　　刷 / 保定市中画美凯印刷有限公司
开　　　本 / 710 毫米 × 1000 毫米　1/16
印　　　张 / 12. 25　　　　　　　　　　　　　　责任编辑 / 多海鹏
字　　　数 / 162 千字　　　　　　　　　　　　　文案编辑 / 把明宇
版　　　次 / 2022 年 4 月第 1 版　2022 年 4 月第 1 次印刷　　责任校对 / 刘亚男
定　　　价 / 79. 00 元　　　　　　　　　　　　　责任印制 / 李志强

编 委 会

前　言

　　本书依据《陆军航空兵学院人才培养大纲》（外训直升机机务维修专业人才培养方案）编写，可用于外国军事留学生（航电维修方向）的教学训练。

　　本书综合考虑受训学员特点，重点关注机务维修专业学员的职业需求，注重理论与实践相结合，图文并茂，以检查站位为主线，介绍了直-9直升机各站位航电设备的主要知识、检查顺序和检查要点，旨在帮助学员巩固理论知识，指导学员完成维护实践操作，提高岗位履职能力。主要内容包括直-9直升机航电设备的工作原理、控制关系、主要参数及维护要求等。本书也可作为该机型维护人员的参考资料。

　　全书共分为三章，第一章为外部检查，第二章为通电检查，第三章为专项检查，由陆航学院外训教、译员编写。感谢直升机使用单位对本书编写给予的配合与指导，同时感谢陆航学院外训大队以及航电与兵器工程系的大力支持。

　　由于水平有限，书中难免有疏误之处，竭诚欢迎读者批评指正！

编　者

2021.05

Preface

Compiled according to *Personnel Training Program of Army Aviation Institute* (*Personnel Training Program for Foreign Military Helicopter Mechanics*), this book can be used to teach and train foreign military trainees (avionic maintenance).

The book takes into consideration the features of foreign military trainees and puts emphasis on the vocational needs of mechanics. Beautifully illustrated and accurately explained, the book combines theory with practice. It mainly focuses on the check on all stations of Z-9 helicopters and introduces the main principles, sequence and key points in checking avionic devices in different stations, so that the trainees could consolidate theories, tamp practical skills and better fulfill their jobs. The main contents of this book include basic working principles, linking and controlling relations, main parameters and maintaining requirements of avionic devices on Z-9 helicopters. The book can also be used as a reference for mechanics of this type of helicopter.

The book falls into three chapters, summarizing the key points in exterior check, power-on check and special check on the helicopter. It is compiled by the instructors and interpreters for the foreign military training of Army Aviation Institute. We'd like to send our thanks to all units that use this type of helicopter for their guidance and cooperation. Our sincere thanks also go to the International Training Center and Department of Avionics and Armament Engineering for their great support.

For any mistakes and shortfalls found in this book, all suggestions and corrections are welcome.

Editors
2021.05

目　录

Contents

直升机检查总体要求与标准

1. 状态（外观）

目视检查（如变形、破损、裂痕、划伤、划痕、腐蚀、清洁、过热或磨损痕迹等）。

2. 固定

目视检查安装点（如锁紧装置状态、自锁螺母的标记等）状态，确保固定牢固、保险规范、减震完好。可辅以触觉检查（即手动检查可能有缺陷的部件），可用手扳动相关部件，通过往复运动能够检测到"不正常"间隙或空隙。

3. 线缆插头

（1）清洁，防磨、绝缘、防油、防水包扎完好，防波套完好且接地良好；

（2）电缆束排列整齐，固定牢靠，绝缘良好，与机件、隔框应有防磨包扎，导线弯曲度要大于90°，活动线缆要留有适当的长度；

（3）插销的安装正确，螺帽拧紧，保险完好、规范；

（4）插钉、插孔、负线等电接触部位要保持清洁、接触良好，搭铁线断丝不应超过1/3。

4. 管路

目视检查，如套管连接处的渗漏及接头固定卡箍的卡紧情况，管路不应有硬折和磨损。注意：重点检查液压、燃油、滑油系统的排泄口、管接头、全静压管等。

使用说明

（1）手册分为外部检查、通电检查和专项检查三大部分。外部检查部分分为 11 个站位。通电检查部分，以各设备通电检查的要求为依据，结合部队工作实践编制而成。

（2）各站位按照内容多少和设备的位置分为若干张照片展示。照片上标明了各设备的名称，设备的检查顺序按照名称前的序号依次进行。

（3）设备名称用白色字体和白色线框标示的，表示该设备不在此处检查。用其他颜色标示的表示需在该站位检查。

（4）在每张设备照片之后附有文字说明，注明各设备的基本原理、参数和检查要求等，设备的排列序号与照片中一致，并且在设备名称后面列出了型号和电气符号，便于进行故障排除和线路查阅（发动机上的设备由于作为一个整体，没有单独电气符号）。

（5）为了减少篇幅，在每条检查项目的最后，列出了该设备的检查要点。各项要点的具体标准集中描述于直升机检查总体要求与标准。

第一章　外部检查

直 -9 直升机检查站位图

站位 1：驾驶舱外部　　　　　　　站位 2：机身左侧

站位 3：主减速器左侧　　　　　　站位 4：发动机左侧

站位 5：尾梁左侧和尾桨毂　　　　站位 6：尾梁右侧和尾减速器

站位 7：机身右侧和底部结构　　　站位 8：主减速器右侧

站位 9：发动机右侧　　　　　　　站位 10：驾驶舱和座舱内部

站位 11：行李舱内部

站位 1

站位 1：驾驶舱外部——机头右侧

（1）卫星定位接收天线 GNSS（43S）：接收卫星定位信号送给 GNSS 接收机（惯性测量部件内）。

检查要点：外观、安装。

（2）速度矢量传感器 GSS−1A（102F）：感受合成气流的全压、静压、总温及相对直升机坐标的方位转角，送入到大气数据计算机中。

检查要点：外观、安装，活动灵活。

（3）罗盘组合天线 ZT−9（81R）：接收来自信标台（或无线电发射台、导航台）的信号作为 WL−9−1 接收机的输入信号。

检查要点：外观、安装。

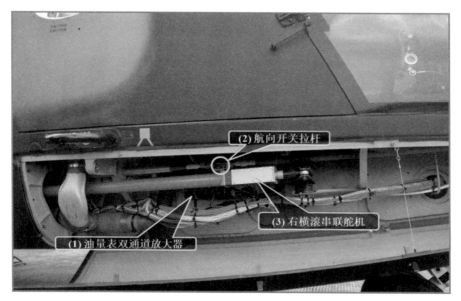

X2000 框前右侧地板下

（1）油量表双通道放大器 FUC-46（2Q）：是 BUC-46 型电容式燃油油量表的组成部分，将油量传感器的电容量变化进行处理，输出与油量成正比的直流电压给指示器。

检查要点：外观、安装、插头、电缆。

（2）航向开关拉杆 CHG-1（28C）：作为硬式连杆装置，将驾驶员施加在脚蹬上的操纵力通过舵机传递给尾桨，使直升机的航向随驾驶员的操纵而变化；作为电气开关，在有自动驾驶仪介入的飞行操纵中，它随着驾驶员操纵航向通道相应开关的通断及自身受力的大小，控制自动驾驶仪系统相应的电气转换（取消锚定），影响不同的飞行操纵状态。

检查要点：外观、安装、插头、电缆。

（3）右横滚串联舵机 DCD-10（24C）：接收放大器的信号，并执行伸出、缩回控制，同时返回舵机位置信号。

检查要点：外观、安装、插头、电缆。

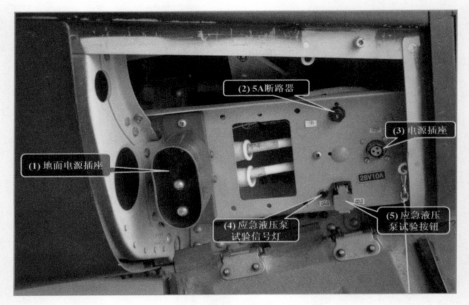

地面电源插座

（1）地面电源插座（23P）：连接地面直流电源，向机上直流系统供电。

检查要点：外观、安装、电缆。

（2）5A断路器（11P）：连接地面电源过压保护插件，当过压保护电路发生短路时断开，地面电源则无法接入电网。

检查要点：外观、安装。

（3）电源插座（90P）：备用直流电插座。

检查要点：外观、安装。

（4）应急液压泵试验信号灯（24G）：地面试验应急液压泵的工作情况时，该灯亮表示工作正常。

检查要点：外观、安装。

（5）应急液压泵试验按钮（20G）：地面试验应急液压泵的工作情况。

检查要点：外观、安装，活动灵活。

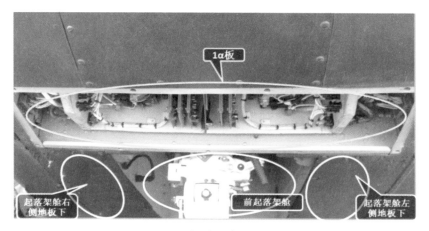

机头下部

1α 板：主配电盒，装有 4 个接触器、2 个短路保护插件、2 个反流保护器、2 个二极管插件、2 个反流割断器、2 个（左、右系统各一）卸载接触器、2 个地面电源继电器、2 个短路探测继电器、1 个地面电源过压保护插件。

检查要点：外观、安装。

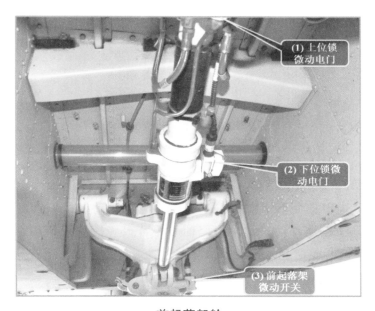

前起落架舱

（1）上位锁微动电门（6G）：起落架收好时锁定，断开收起电路；起落架在收放和放好时开锁，准备好收起电路。

检查要点：外观、安装、插头、电缆。

（2）下位锁微动电门（7G）：起落架放好时锁定，断开放下电路；起落架在收放和收好时开锁，准备好放下电路。

检查要点：外观、安装、插头、电缆。

（3）前起落架微动开关（5G）：缓冲支柱微动开关，当缓冲支柱压缩（地面停放时）且中立时接通，防止收起起落架。

检查要点：外观、安装、插头、电缆。

前起落架舱左侧地板下

（1）火警控制盒 HKH−8A（7E）：当左发动机区域温度超过预定值时，继电器动作，发出报警信号。若火警探测回路发生断路或接地信号时，控制盒发出故障信号。

检查要点：外观、安装、插头、电缆。

（2）接线盒（30α）：包括液压系统、通信系统、防冰排雨系统、起落架、照明系统、配电以及发动机控制系统等的接线。

检查要点：外观、安装、插头、电缆。

（3）继电器盒（23α）：包括防火系统、液压系统、舱门告警和起落架收放等系统的继电器。

检查要点：外观、安装、插头、电缆。

（4）插头（48A）：左右总距杆的操纵电位计和应急抛放电路转接插头。

检查要点：外观、安装、插头、电缆。

（5）接线盒（31α）：照明系统和配电设备的接线盒。

检查要点：外观、安装、插头、电缆。

前起落架舱右侧地板下

（1）火警控制盒 HKH-8A（6E）：当右发动机区域温度超过预定值时，继电器动作，发出报警信号。若火警探测回路发生断路或接地信号时，控制盒发出故障信号。

检查要点：外观、安装、插头、电缆。

（2）继电器盒（24α）：包括防火、液压系统、防冰排雨系统、照明系统等的控制继电器。

检查要点：外观、安装、插头、电缆。

（3）插头（14α）：正驾驶周期变距杆插头。

检查要点：外观、安装、插头、电缆。

（4）侧向加速度计 GJ−10B（58F）：用于感受飞机的侧向加速度，并将其变化转换成为 400Hz 信号输入给驾驶仪的航向通道，实现飞机的协调转弯。

检查要点：外观、安装、插头、电缆。

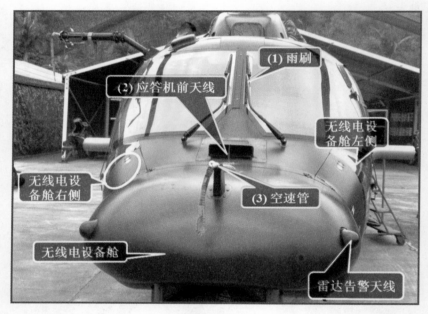

机头前部

（1）雨刷：刮除风挡玻璃上的雨雪。快速不少于 60 次 / 分，慢速不少于 40 次 / 分，机械角最大 45°，直流 27V，电机功率 141W，不允许在干燥风挡玻璃上工作。

检查要点：外观、安装、刷片、压力。

（2）应答机前天线（72S）：接收地面、舰载或其他机载询问机（或询问应答机）的识别询问信号（一般 / 编 1/ 编 2/ 编 3/ 编 4）、非识别询问信号（批号、高度），送给应答机。

检查要点：外观、安装。

（3）空速管：感受全压信号，传给空速表和大气数据计算器。

检查要点：外观、安装、管路。

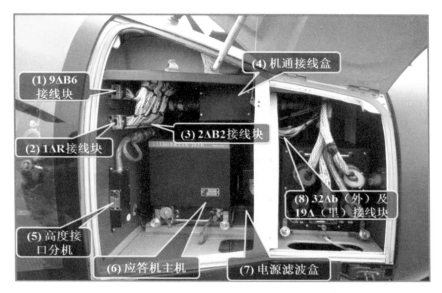

（1）9ΔB6
接线块

（4）机通接线盒

（2）1ΔR接线块

（3）2ΔB2接线块

（5）高度接
口分机

（8）32Δb（外）及
19Δ（里）接线块

（6）应答机主机

（7）电源滤波盒

无线电设备舱右侧

（1）（2）（3）（8）接线块：9ΔB6：导航系统和控制增稳系统接线块。1ΔR：机内通话器接线排。2ΔB2：电源系统和发动机控制系统接线块。32Δb：控制增稳、燃油液压和指示系统接线盒。19Δ：防冰排雨系统接线盒。

检查要点：外观、安装、电缆。

（4）机通接线盒（22R）：内装有接收、发射、放大以及电源滤波电路和告警电路。插座 J1、J2 和 J3 连接到控制盒。J6 与机上无线电设备以及电源相连。J4 和 J5 不用。

检查要点：外观、安装、插头、电缆。

（5）高度接口分机 9A1（79S）：可接收来自高度表或大气数据系统的未经修正的压力高度。具有 ARINC-429 大气数据系统接口和 RS-422 高度接口。

检查要点：外观、安装、插头、电缆。

（6）应答机主机 K/LKB01 5A（70S）：可对地面、舰载或其他机载询问机（或询问应答机）的识别询问信号（一般／编 1／编 2／编 3／编 4）、非识别询问信号（批号、高度）接收译码并发射相应的识别应答信号、非识别应答信号（批号、高度）或特殊应答信号（特 1～特

5），使询问机能根据应答信号判断载机的目标属性和录取载机的批号和高度信息。

检查要点：外观、安装、插头、电缆。

（7）电源滤波盒 25D（75S）：对电源信号进行滤波。

检查要点：外观、安装、插头、电缆。

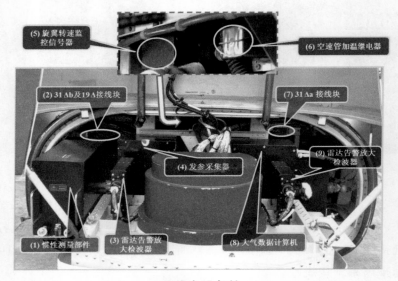

无线电设备舱

（1）惯性测量部件 HJG-1A（41S）：接收加速度及陀螺信号，经处理后输出显示信息及武器、导航信息。

检查要点：外观、安装、插头、电缆。

（2）31Δb 及 19Δ 接线块：防水防雨系统接线块。

检查要点：外观、安装、电缆。

（3）（9）雷达告警放大检波器 ARW9503-1（5T、8T）：接收天线的雷达信号，经放大与检波送给信号分析器。

检查要点：外观、安装、插头、电缆。

（4）发参采集器 EFC-2-02（204E）：产生±15V 直流电源供各有关传感器工作使用（主减、液压、发动机），从各有关传感器及设备上得到所需的信息（主减、液压、发动机的压力、温度、转速、扭矩、状态），接收轮载信息、电源参数、环境温度和飞行高度信息。将所接收到的信息进行计算、处理，然后输出到相关设备（发参显示器与告警系统）。

检查要点：外观、安装、插头、电缆。

（5）旋翼转速监控信号器 XJZ-6（11E）：监控旋翼转速（与832E 配套），并发出高音低音报警信号。

检查要点：外观、安装、插头、电缆。

（6）空速管加温继电器（4F）：控制空速管加温电路的接通与断开。

检查要点：外观、安装、电缆。

（7）31Δa 接线块：接线块。

检查要点：外观、安装、电缆。

（8）大气数据计算机 XSC-5A（101F）：与综合任务系统配套，通过接收速度矢量传感器感受合成气流的全压、静压、总温及相对直升机坐标的方位转角，同时接收总线送来的垂直过载、旋转角速度、直升机重量等有关信息，经过处理、解算得到直升机所需的三轴空速、气压高度、升降速度、总温、旋翼诱导空速、侧滑角、攻角等参数。系统具有自检测功能。

检查要点：外观、安装、插头、电缆。

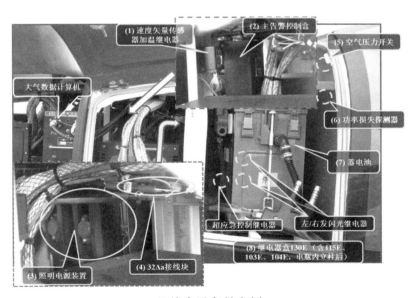

无线电设备舱左侧

（1）速度矢量传感器加温继电器 JKA–52B（106F）：控制加温电路的通断。

检查要点：外观、安装、电缆。

（2）主告警控制盒 KZH–55A（11α）：根据传感器信号接通相应的告警灯，并为主告警灯提供闪光电源。

检查要点：外观、安装、插头、电缆。

（3）照明电源装置（50L）：为座舱照明设备提供电源。

检查要点：外观、安装、插头、电缆。

（4）32Δa 接线块：增稳、防火、燃油、液压、指示、起落架转接模块。

检查要点：外观、安装、电缆。

（5）空气压力开关（4G）：与起落架警告灯（装于驾驶员侧仪表板上）相连，当飞行速度相当于或低于 101.92km/h(55kn) 时，起落架尚未放下锁死时提供告警信号。

检查要点：外观、安装、插头、电缆。

（6）功率损失探测器 XJZ–2A（32E）：感受转速传感器的频率信号（Nf），当达到校准临界值时，探测器接地，使有关的功率警告（损失）灯燃亮。探测临界值：转子转速 313r/min。精确度：±1%。从探测临界值到发动机停车有警告信号输出（警告灯燃亮）。

检查要点：外观、安装、电缆。

（7）蓄电池（22P）：应急和备用直流电源，额定容量 36Ah，额定电压 24V。

检查要点：外观、安装、插头、电缆。

（8）继电器盒（130E 含 115E、103E、104E）：如果有一台发动机发生故障，双发扭矩差 ΔC>25% 时，接通电路使超应急功率警告灯燃亮（发参上边内侧）。

检查要点：外观、安装、插头、电缆。

（1）俯仰位置传感器 GE–23（71C）：感受纵向角位置信号并传给计算机。

检查要点：外观、安装、插头、电缆。

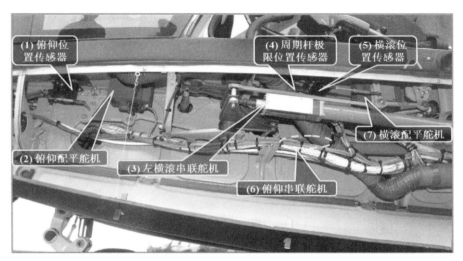

X2000 框前左侧地板下

（2）俯仰配平舵机 DCD-9（21C）：接收计算机指令，给操纵杆系建立一个锚定点。有阻尼器，当电磁离合器部件不啮合（不通电）时，舵机被锚定。装有载荷敏感开关，当驾驶杆有操纵力矩时，杆力被力释放。

检查要点：外观、安装、插头、电缆。

（3）左横滚串联舵机 DCD-10B（25C）：接收放大器的信号，并执行伸出、缩回控制，同时返回舵机位置信号。

检查要点：外观、安装、插头、电缆。

（4）周期杆极限位置传感器：周期杆左右极限位置时，极限灯亮。

检查要点：外观、安装、插头、电缆。

（5）横滚位置传感器 GE-23（73C）：感受横向角位置信号并传给计算机。

检查要点：外观、安装、插头、电缆。

（6）俯仰串联舵机 DCD-10（23C）：接收放大器的信号，并执行伸出、缩回控制，同时返回舵机位置信号。

检查要点：外观、安装、电缆。

（7）横滚配平舵机 DCD-9（27C）：接收计算机指令，给操纵杆系建立一个锚定点。有阻尼器，当电磁离合器部件不啮合（不通电）

时，舵机被锚定。装有载荷敏感开关，当驾驶杆有操纵力矩时，杆力被力释放。

检查要点：外观、安装、插头、电缆。

站位 2

站位 2：机身左侧

（1）静压孔：感受空气静压传给全静压系统和大气数据计算器。

检查要点：外观、畅通性。

应急液压舱

（1）辅助液压盒（接线盒）（19G）：辅助液压系统接线盒。

检查要点：外观、安装、插头、电缆。

（2）应急液压电磁阀 YDK-12（26G）：通电时打开应急液压系统管路。

检查要点：外观、安装、插头、电缆、渗漏。

（3）应急液压开关 YCG-3（23G）：当应急液压力超过 95bar 时，灯亮，表示应急系统工作。

检查要点：外观、安装、插头、电缆、渗漏。

（4）应急液压泵 YCB-0.25（21G）：在主液压系统出现故障的情况下，向主系统和辅助系统提供液压能源。

检查要点：外观、安装、插头、电缆、渗漏。

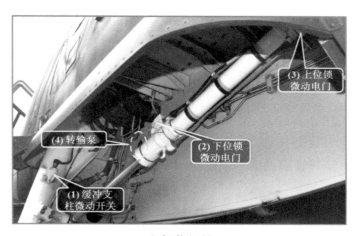

左起落架舱

（1）缓冲支柱微动开关（9G）：感受起落架离地信号，防止缓冲支柱未完全伸长（离地）时，起落架收起。

检查要点：外观、安装、插头、电缆。

（2）下位锁微动电门（13G）：当起落架完全放下时，断开液压电磁活门电路，同时接通收起预备电路。

检查要点：外观、安装、插头、电缆。

（3）上位锁微动电门（11G）：当起落架完全收起时，断开液压电磁活门电路，同时接通放下预备电路。

检查要点：外观、安装、插头、电缆。

（4）转输泵RXB-4（20Q）：将燃油从一个油箱组转输到另一个油箱组。

检查要点：外观、安装、插头、电缆、渗漏。

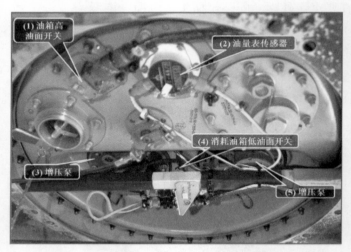

机身底部（左）（右侧与左侧相同，但没有高油面开关）

（1）油箱高油面开关XU-24（9Q）：在燃油转输期间，当相应的燃油箱转输满油时，高油面开关电路接通并且相应的"高油面（HI.LEV）"指示灯燃亮。

检查要点：外观、安装、插头、电缆、渗漏。

（2）油量表传感器GUC-46/3（5Q）：是电容式油量测量系统部件，提供有关油箱内贮存的油量数据。

检查要点：外观、安装、插头、电缆、渗漏。

（3）（5）增压泵RLB-12（7Q、69Q）：提高燃油管路压力。

检查要点：外观、安装、插头、电缆、渗漏。

（4）消耗油箱低油面开关XU-24/3（11Q）：当消耗油箱的燃油低于18L时，接通电路，在3Q控制板上消耗油箱低油面"FUEL.Q"警告灯燃亮。在信号板7α相应的"FUEL.Q"警告灯燃亮。操纵警告系统逻辑电路11α，至此在仪表板上"主警告（ALARM）"灯闪亮。

检查要点：外观、安装、插头、电缆、渗漏。

站位 3

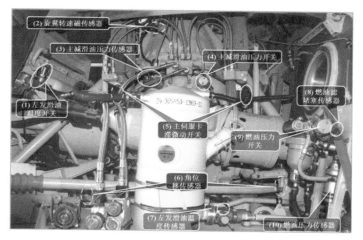

站位 3：主减速器左侧

（1）左发滑油温度开关 TW-9（47E）：感受左发动机滑油温度，130℃时告警灯亮。

检查要点：外观、安装、插头、电缆、渗漏。

（2）旋翼转速磁传感器 GCZ-3（832E）：感受翼转速信号，传给音响报警信号器及 7α 板信号灯。低音报警是 85Hz~172.5Hz（旋翼转速为 170r/min~345r/min）；高音报警是 190Hz（旋翼转速为大于 380r/min）。

检查要点：外观、安装、插头、电缆。

（3）主减滑油压力传感器 CY-YZ-0101-0.3（17E）：用来测量主减速器润滑系统的滑油压力。与发参采集器相连。

检查要点：外观、安装、插头、电缆、渗漏。

（4）主减滑油压力开关（16E）：当滑油压力降到 0.08MPa 时引起警告灯燃亮。

检查要点：外观、安装、插头、电缆。

（5）主伺服卡滞微动开关：转阀故障告警装置，通过一个"助力器卡滞"警告灯给驾驶员告警信号。

检查要点：外观、安装、插头、电缆。

（6）角位移传感器：感受操纵系统位置信号，传给飞行参数系统。加装设备无电气符号。

检查要点：外观、安装、电缆。

（7）左发滑油温度传感器 GWR-2A（21E）：感受左发动机滑油温度，转变成电信号传递到发动机参数采集显示系统进行处理并显示。

检查要点：外观、安装、插头、电缆、渗漏。

（8）燃油滤堵塞传感器（18Q）：压力差 140±20mbar 时，3Q 板上的"油滤（FILT）"堵塞指示灯燃亮。信号板 7α 上的"燃油（FUEL）"指示灯燃亮。

检查要点：外观、安装、插头、电缆、渗漏。

（9）燃油压力开关（14Q）：感受低压泵出口压力，当低于 0.2bar 时，给座舱低压指示。7α 上的"燃油（FUEL）"灯燃亮。3Q 板上的"燃油压力（FUEL.P）"灯燃亮。

检查要点：外观、安装、插头、电缆、渗漏。

（10）燃油压力传感器 CY-YZ-0101-0.2（17Q）：传感器的阻值是根据燃油压力变化而发出相应电压信号，经过发参采集器处理后，送到发参显示器显示。测量范围（0~0.2）MPa。

检查要点：外观、安装、插头、电缆、渗漏。

站位 4

站位 4：发动机左侧（左发动机舱）

（1）超应急电磁活门：飞行中，如一台发动机故障，双发扭矩差ΔC>25%时，超应急电磁活门接通，从而允许获得超应急供油，同时超应急功率警告灯燃亮（发参上边内侧）。

检查要点：外观、安装、插头、电缆、渗漏。

（2）放气活门：发动机起动过程中，打开活门以调节进气量，防止发动机喘振。

检查要点：外观、安装、插头、电缆。

（3）起动电磁活门 RDK-16：喷油电磁活门，起动初期向燃烧室提供燃油。当燃油压力达到 0.24MPa（2.4bar）时，停油喷气。

检查要点：外观、安装、插头、电缆、渗漏。

（4）滑油压力开关 HJK-3：当滑油压力降到 1.3bar 时，"ENG.1"警告灯亮。

检查要点：外观、安装、插头、电缆、渗漏。

（5）滑油压力传感器 CY-YZ-0101-0.8：压阻式压力传感器包括与感应线圈相连的压力膜盒，测量每台发动机的滑油压力，压力测量范围（0~0.8MPa），传送到发动机参数采集显示系统进行处理并显示。

检查要点：外观、安装、插头、电缆、渗漏。

（6）点火电嘴：将油－气混合气点燃。

检查要点：外观、安装、电缆。

（7）金属屑探测器：监测发动机轴承腔内的磨损情况，若磁性屑末被吸附接通环形间隙，座舱中的指示灯发光，给飞行员报警。

检查要点：外观、安装、插头、电缆、渗漏。

（8）Nf 传感器 GZ-20：检测发动机自由涡轮转速，发送给发参采集器，用于发动机功率损失告警。

检查要点：外观、安装、插头、电缆、渗漏。

（9）超转停车电磁活门：自由涡轮转速超过 123.1%＋2.5%额定转速，两个传感器感受超转，电磁阀使发动机停车。

检查要点：外观、安装、插头、电缆、渗漏。

（10）超转传感器：感受自由涡轮转速信号，进行超转报警和超转停车控制。

检查要点：外观、安装、插头、电缆、渗漏。

站位 5

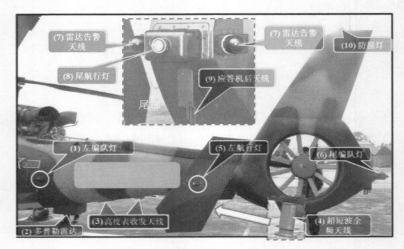

站位 5：尾梁左侧

（1）（6）左编队灯 BD-12（147L）、尾编队灯 BD-12（144L）：用于提供机群编队飞行的其他直升机飞行员所要求的信息，有"红外""正常"两种模式。

检查要点：外观、安装。

（2）多普勒雷达 777（22S）：利用多普勒效应测定目标的位置及速度的雷达。

检查要点：外观、安装。

（3）高度表收（前、13S）发（后、14S）天线：发射天线向地面发调频连续波信号。接收天线发射的回波信号。

检查要点：外观、安装。

（4）超短波全频天线 TKR128（43R）：超短波电台全频段天线。工作频率为 30MHz~87.975MHz 和 108MHz~155.975MHz 等。主要用于直升机与直升机之间、直升机与地面之间明话／密话通信联络。

检查要点：外观、安装。

（5）（8）左航行灯 HD-13（15L）、尾航行灯 WD-3A（16L）：左红右绿（4.2W）尾白（2.4W），用于确定飞行中直升机的位置和飞行方向,有正常和隐蔽两种模式。

检查要点:外观、安装。

(7)雷达告警天线(左3S、右4S):截获雷达的探测信号,传递给双侧向接收机进行处理。

检查要点:外观、安装。

(9)应答机后天线(73S):可对地面、舰载或其他机载询问机(或询问应答机)的识别询问信号(一般/编1/编2/编3/编4)、非识别询问信号(批号、高度)接收译码并发射相应的识别应答信号、非识别应答信号(批号、高度)或特殊应答信号(特1~特5),使询问机能判断载机的目标属性和录取载机的批号和高度信息。

检查要点:外观、安装。

(10)防撞灯FZD-20(18L):正常和隐蔽模式。正常模式下可以确定在远距离飞行和地面运行中直升机的位置。隐蔽模式下用来在远距离飞行时,被空中或地面运动中的直升机发现,以避免相撞。28V,60次/分。

检查要点:外观、安装。

站位6

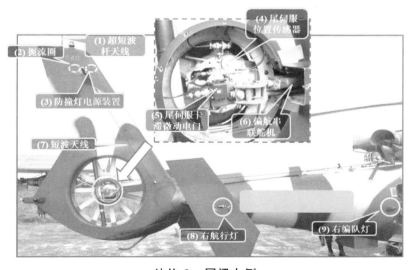

站位6:尾梁右侧

（1）超短波杆天线（52R）：接收射频信号，发射载波信号。频段 30MHz~87.975MHz。

检查要点：外观、安装。

（2）扼流圈（97L）、（3）防撞灯电源装置 KZH–125（13L）：为防撞灯提供电源及闪光触发信号，使之发出较强的红色频闪信号，以避免与其他飞机或航空飞行器相撞。

检查要点：外观、安装、插头、电缆。

（4）尾伺服位置传感器（30FA）：向飞行参数系统传送尾伺服位置信号。

检查要点：外观、安装、插头、电缆。

（5）尾伺服卡滞微动电门（11D）：转阀故障告警装置，通过一个"助力器卡滞"警告灯给驾驶员告警信号。

检查要点：外观、安装、插头、电缆。

（6）偏航串联舵机 DCD–10A（26C）：接收放大器的信号，并执行伸出、缩回控制，同时返回舵机位置信号。

检查要点：外观、安装、插头、电缆。

（7）短波（钢索）天线（13R）：接收短波波段射频信号，发射载波信号。

检查要点：外观、安装。

（8）右航行灯 HD–13（17L）、（9）右编队灯 BD–12（146L）：功能和参数同左侧。

检查要点：外观、安装。

站位 7

（1）超短波天线（55R）：接收射频信号，发射载波信号。波段 108MHz~155.975MHz、156MHz~173.975MHz、225MHz~399.975MHz。

检查要点：外观、安装。

（2）静压孔：测量大气静压，传给气压高度表和大气数据计算器。

检查要点：外观、畅通。

（3）总距位置传感器 GE–14（66C）：感受总距通道的操纵位置信号并传给计算机。

站位 7：机身右侧

检查要点：外观、安装、插头、电缆。

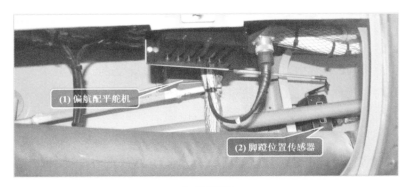

右短翼下

（1）偏航配平舵机 DCD-9A（22C）：并联舵机。接收放大器指令，给操纵杆系建立一个锚定点。无阻尼器，当电磁离合器部件啮合（通电）时，舵机被锚定。

检查要点：外观、安装、插头、电缆。

（2）脚蹬位置传感器 GE-15（29C）：感受航向通道的操纵位置信号并传给计算机。

检查要点：外观、畅通、插头、电缆。

站位 8

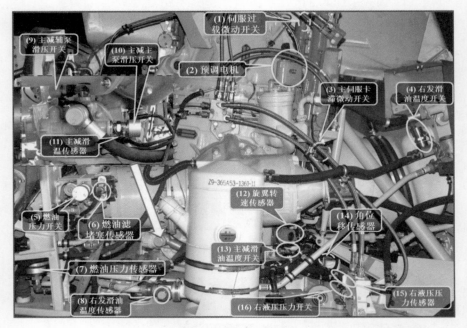

站位 8：主减速器右侧

（1）伺服过载微动开关：检测伺服机构承受的最大载荷，"极限 (LIMIT)"指示灯燃亮。可通过按钮对电路进行试验。

检查要点：外观、安装、插头、电缆。

（2）预调电机：用于改善发动机对总桨距变化的反应时间，预先进行速度调节。

检查要点：外观、畅通、插头、电缆。

（3）主伺服卡滞微动开关（3D）：探测主伺服控制分配阀卡死，使信号板 7α 上的"伺服 (SERVO)"警告灯燃亮和"主警告 (ALARM)"灯闪亮。6α 板上有一个试验开关可以用来检查系统无液压状态是否正确工作。

检查要点：外观、安装、插头、电缆。

（4）右发滑油温度开关（40E）：当温度达到 125℃时，开关触点闭合，信号板 7α 上的"滑油温度（OIL.TEMP）"警告灯燃亮。

检查要点：外观、安装、插头、电缆、渗漏。

（5）燃油压力开关（14Q）：感受低压泵出口压力，当燃油压力低于 0.2bar 时，7α 上的"燃油（FUEL）"指示灯燃亮。在燃油控制板 3Q 上的"燃油压力（FUEL.P）"指示灯燃亮。

检查要点：外观、安装、插头、电缆、渗漏。

（6）燃油滤堵塞传感器：压力差 140±20mbar 时，燃油控制板 3Q 上的"油滤(FILT)"堵塞指示灯燃亮。信号板 7α 上的"燃油(FUEL)"指示灯燃亮。

检查要点：外观、安装、插头、电缆、渗漏。

（7）燃油压力传感器：传感器的电阻值是根据燃油压力发生变化而发出相应电压信号，经过发动机参数采集器采集处理后，送到仪表板上发动机参数显示器（202E）上显示。测量范围（0~0.2）MPa。

检查要点：外观、安装、插头、电缆、渗漏。

（8）右发滑油温度传感器（22E）：根据镍丝电阻系数随着温度的变化而工作，传感器与发参系统采集器相连接。

检查要点：外观、安装、插头、电缆、渗漏。

（9）主减辅泵滑压开关 YKJ-4004-0.8（66E）、（10）主减主泵滑压开关 YKJ-4004-0.8（65E）：主泵压力开关与"主减主泵（MAIN.PMP）"灯相连。辅泵压力开关，与信号板上"辅助泵（AUX.PMP）"警告灯相连。（小于 0.6bar 亮）。

检查要点：外观、安装、插头、电缆、渗漏。

（11）主减滑温传感器（13E）：利用镍导线电阻随温度变化而变化的特性，测量安装处的滑油温度。与发参采集器相连。

检查要点：外观、畅通、插头、电缆、渗漏。

（12）旋翼转速传感器（15E）：提供交流电压频率信号给发参显示器，此频率正比于音轮的转速。

检查要点：外观、安装、插头、电缆、渗漏。

（13）主减滑油温度开关（14E）：当温度达到 132.5±2.5℃时，开关触点闭合，信号板 7α 上的"滑油温度（OIL.TEMP）"警告灯燃亮。

检查要点：外观、安装、插头、电缆、渗漏。

（14）角位移传感器：感受右前助力器位置信号，为飞参系统专用传感器。

检查要点：外观、安装、插头、电缆。

（15）右液压压力传感器 CY-YZ-0101-10（4D）：用来提供主液压系统的压力信号（0-10MPa）。

检查要点：外观、安装、插头、电缆、渗漏。

（16）右液压压力开关 CY-YCG-4（6D）：感受油压，并在预先调定的升压和降压值时接通电路。升压 2.5±0.5MPa 时灭，降压 1+0.5MPa 时亮。

检查要点：外观、安装、插头、电缆、渗漏。

站位 9

站位 9：发动机右侧（右发动机舱）

（1）起动电磁活门 RDK-16：喷油电磁活门，起动初期向燃烧室提供燃油。当燃油压力达到 0.24MPa（2.4bar）时，停油喷气。

检查要点：外观、安装、插头、电缆、渗漏。

（2）放气活门 ZQK-2：在发动机起动时，控制（气动控制）放气活门打开（P_2/P_0 小于一定值时），调节供气量防止喘振，在

90%~100% Ng（P_2/P_0 大于一定值时）之间关闭。

检查要点：外观、安装、插头、电缆、渗漏。

（3）点火电嘴 BDH-15：半导体电嘴，起动期间点燃燃烧室混合气体。

检查要点：外观、安装、插头、电缆。

（4）高能点火器 DHZ-20B：电容式，将低压直流电变换成高压脉冲电（2 000V），供点火电嘴产生火花。

检查要点：外观、安装、插头、电缆。

（5）超转传感器 GZ-18：起动时，Nf 达到约 24％时"超转（O/SPEED）"指示灯熄灭，监测系统工作。Nf 超过 123.1％+2.5％时，传感器感受超转，安全电磁阀使发动机停车。正常停车引起"超转（O/SPEED）"指示灯再次燃亮。如果灯熄灭，必须按下超转探测盒上的复位按钮，使指示灯燃亮。

检查要点：外观、安装、插头、电缆。

（6）热电偶 GR-14：检测发动机排气口处的排气温度，并将检测到的信息传递到发动机参数采集显示系统进行处理并显示。

检查要点：外观、安装、插头、电缆。

（7）Ng 传感器 GZ-17：检测发动机燃气涡轮转速，发送给发参采集器，用于监控发动机状态。

检查要点：外观、安装、插头、电缆。

（8）起动发电机 QF-4.8：起动时，作为电动机带动发动机转动，起动完毕后作为发电机供机上直流电源。

检查要点：外观、安装、插头、电缆。

（9）扭矩传感器 GY-20：检测发动机输出功率。

检查要点：外观、安装、插头、电缆、渗漏。

（10）火警传感器 JUI-18：感受发动机火警信号，临界值为200℃。

检查要点：外观、安装、电缆。

站位 10

应急磁罗盘

7α板

多功能彩色显示器

多功能彩色显示器

应急地平仪

发参显示器

地平仪

场压给定器

机箱控制盒

油量表

起落架信号灯

耦合器操纵台

航空时钟

航向位置指示器

起落架操纵开关

多功能键盘

航行灯编队灯操纵台

站位 10：驾驶舱——仪表板

多功能键盘

航行灯编队灯操纵台

雨刷电机

防撞灯操纵台

增稳控制台

燃油控制板

着陆灯开关

舱音记录仪面板

着陆灯收放开关

5α板

无线电罗盘控制盒

超短波电台I控制盒

超短波电台II控制盒

短波电台控制盒

干扰弹控制盒

机载应答机控制盒

空管应答机控制盒

操纵台

驾驶舱顶棚

（1）12α 板：电气控制板。

检查要点：外观、安装。

（2）（4）活动灯 ZCD-16（19L、30L）：读图和局部照明。

检查要点：外观、安装、插头、电缆。

（3）报警喇叭（20E）：用两种音响对机组成员进行报警，高音调信号用于旋翼超速警报，低音调信号用于旋翼低速警报。低音调报警是旋翼转速为 170r/min~345r/min；高音调报警是旋翼转速为大于 380r/min。

检查要点：外观、安装、插头、电缆。

（5）顶灯 DD-7A（22L）：座舱环境照明。

检查要点：外观、安装、插头、电缆。

（6）备用蓄电池 CH-130（124L）：用于直升机事故时，顶灯和指令灯的应急照明电源。

检查要点：外观、安装、插头、电缆。

（7）20Δ 接线块、（8）815ΔK 接线块：通信系统、照明系统、主减滑油、发动机监测、发动机起动系统接线。

检查要点：外观、安装、电缆。

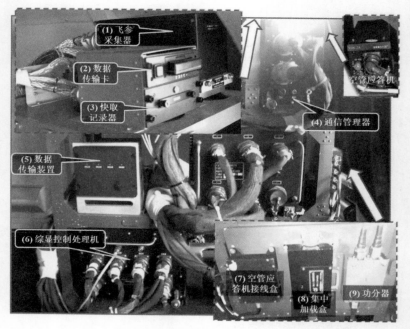

座舱设备架

（1）飞参采集器：用于采集处理飞行参数系统的数据。

检查要点：外观、安装、插头、电缆。

（2）数据传输卡：用来存储向系统装订的战术数据及其他参数，具有应急毁钥能力。

检查要点：外观、安装、插头、电缆。

（3）快取记录器：能够与防毁记录器一起记录部分飞参数据，用于快速卸载飞参数据。

检查要点：外观、安装、插头、电缆。

（4）通信管理器 CH−130 (124L)：载机平台数据链控制与管理、消息处理的核心。与航电系统和超短波电台、短波电台共同实现载机平台的数据链系统的功能。

检查要点：外观、安装、插头、电缆。

（5）数据传输装置：用来向系统装订战术数据及其他参数。可加载航路点、航线计划、外挂武器种类、数量、直升机起飞时重量等信息。

检查要点：外观、安装、插头、电缆。

（6）综显控制处理机 XKG–9B（7U）：DCP，是综合显示控制系统的核心部件。它接受各有关任务设备输出的数据或信号，进行综合处理，从而在多功能彩色显示器上显示有关飞行参数和攻击画面。

检查要点：外观、安装、插头、电缆。

（7）空管应答机接线盒：接线盒。

检查要点：外观、安装、插头、电缆。

（8）集中加载盒（101R）：用于 CMU 数据记录和视频记录，同时还实现 4 个串口和 2 个 USB 口接口物理连接。

检查要点：外观、安装、插头、电缆。

（9）功分器：将应答机主机的信号分配给前后天线。

检查要点：外观、安装、插头、电缆。

站位 11

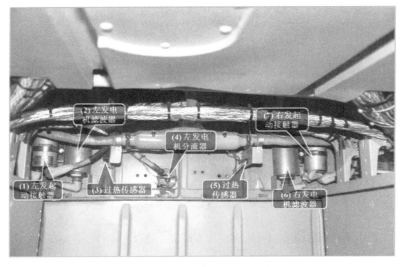

站位 11：行李舱——前部

（1）（7）左、右发起动接触器（25P 和 24P）：用于发动机起动时接通起动发电机起动电路。

检查要点：外观、安装、插头、电缆。

（2）（6）左、右发电机滤波器（27P 和 26P）：发电时滤除交流干扰，起动时通过滤波控制无线电干扰。

　　检查要点：外观、安装、插头、电缆。

　　（3）（5）过热传感器（33W 和 32W）：在行李舱温度达到 110℃ 时，过热探测器的双金属片触点断开，7α 上的"过热 (O/HEAT)"指示灯燃亮。

　　检查要点：外观、安装、插头、电缆。

　　（4）左发电机分流器（28P）：测量发动机输出电流值。额定电流值为 160A。

　　检查要点：外观、安装、插头、电缆。

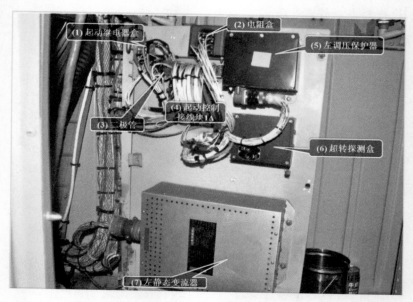

左电气设备安装板 21α

　　（1）起动继电器盒（7K）：发动机起动控制继电器盒。
　　检查要点：外观、安装、插头、电缆。
　　（2）电阻盒：发动机灭火系统电路补偿电阻盒（包含 17W、19W、29W、31W）。
　　检查要点：外观、安装、电缆。
　　（3）二极管（117P）：左调压保护器的二极管。
　　检查要点：外观、安装、电缆。

（4）起动控制接线块 1Δ：左发动机控制与左电源系统控制电路接线块。

检查要点：外观、安装、电缆。

（5）左调压保护器 TKB-3（29P）：监测发电机的工作并调节发电机输出电压。调节（发电状态调节电压，起动状态调节电流）、控制和保护作用。

检查要点：外观、安装、插头、电缆。

（6）超转探测盒（6K）：与超转传感器共同完成超转探测，正常情况下，Ng 小于 25% 时使超转灯亮；超转时（Ng 大于 123.1%）驱动安全阀使发动机停车。它还包括试验按钮和复位按钮。

检查要点：外观、安装、插头、电缆。

（7）左静态变流器 PC-1000（41X）：将直流电变换为交流电，供机上交流用电设备使用。输入 28V 直流电，输出 115V/400Hz、26V/400Hz 交流电。

检查要点：外观、安装、插头、电缆。

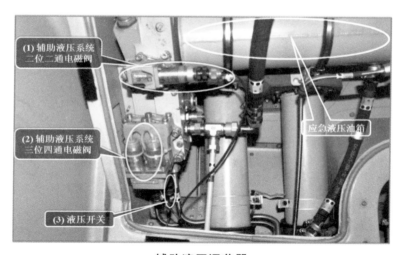

辅助液压调节器

（1）辅助液压系统二位二通电磁阀 YDK-11（15G）：辅助液压系统控制阀，通电时使管路中液压油流回油箱，断电时使管路产生压力。

检查要点：外观、安装、插头、电缆。

（2）辅助液压系统三位四通电磁阀YDK-10（16G）：控制液压油的流向从而控制起落架的收放。

检查要点：外观、安装、插头、电缆。

（3）液压开关YCG-4（22G）：当辅助液压系统压力大于2.5±0.5MPa时，接通，"AUX.HYD"信号灯亮。

检查要点：外观、安装、插头、电缆。

右电气设备安装板22α

（1）右调压保护器TKB-3（28P）：监测发电机的工作并调节发电机输出电压。调节（发电状态调节电压，起动状态调节电流）、控制和保护作用。

检查要点：外观、安装、插头、电缆。

（2）电阻盒：发动机灭火系统电路补偿电阻盒，包含18W、20W、28W、30W 。

检查要点：外观、安装、电缆。

（3）起动控制接线块2Δ：右发动机控制与右电源系统控制电路接线块。

检查要点：外观、安装、电缆。

（4）二极管（118P）：右调压保护器的二极管。

检查要点：外观、安装、电缆。

（5）起动继电器盒（8K）：发动机起动控制继电器盒。

检查要点：外观、安装、插头、电缆。

（6）超转探测盒（5K）：与超转传感器共同完成超转探测，正常情况下，Ng 小于 25% 时使超转灯亮；超转时（Ng 大于 123.1%）驱动安全阀使发动机停车。它还包括试验按钮和复位按钮。

检查要点：外观、安装、插头、电缆。

（7）右静态变流器 PC-1000（40X）：将直流电变换为交流电，供机上交流用电设备使用。输入 28V 直流电，输出 115V/400Hz、26V/400Hz 交流电。

检查要点：外观、安装、插头、电缆。

（8）信号分析器（9T）：在接到放大检波器送来的雷达信号后，经过处理将告警信息送给综合显示控制器以便飞行员进行操作。

检查要点：外观、安装、插头、电缆。

（9）备份航姿处理机（112F）：接收磁航向和无线电航向信号，经处理后送航向指示器，为直升机提供备份航向。

检查要点：外观、安装、插头、电缆。

（10）罗盘接收机：接收组合天线信号，可工作在收讯和罗盘方式，经处理后向显示器输出方位信号和向机内通话输出音频信号。

检查要点：外观、安装、插头、电缆。

（11）超短波电台收发信机 TKR128（10R）：具有收信和发射功能，用于直升机与直升机之间、直升机与地面之间明话 / 密话通信联络。

检查要点：外观、安装、插头、电缆。

（12）电台共址滤波器：降低本机设备发射时的宽余带噪声，有效减少多部共址设备的收发干扰，提高多部通信设备的共址工作能力。

检查要点：外观、安装、插头、电缆。

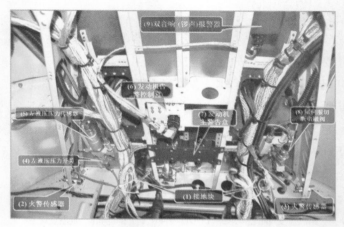

行李舱顶棚（向前）

（1）接地块（15N、16N）：接地线连接点。

检查要点：外观、安装、电缆。

（2）（3）火警传感器（25W、26W）：感受行李舱火警信号，当温度达到160℃时"行李舱火警（CARGO.F）"指示灯亮。

检查要点：外观、安装、电缆。

（4）左液压压力开关YCG–4（5D）：感受油压，并在预先调定的升压和降压值时接通电路。升压2.5±0.5MPa时灭，降压1+0.5MPa时亮。

检查要点：外观、安装、插头、电缆、渗漏。

（5）左液压压力传感器CY–YZ–0101–10（3D）：用来提供主液压系统的压力信号（0~10MPa）。

检查要点：外观、安装、插头、电缆、渗漏。

（6）发动机告警控制盒（815K）：音响告警电源装置，向发动机音响告警设备提供电源。

检查要点：外观、安装、插头、电缆。

（7）发动机主警告盒（810K）：接受发参采集器的信号，经处理产生音响告警信号（送给耳机）。

检查要点：外观、安装、插头、电缆。

（8）尾伺服切断电磁阀YDK–12（12D）：当右液压油箱油面下降到2L时，自动切断尾桨液压供压油路。

检查要点：外观、安装、插头、电缆。

（9）双音响（锣声）报警器 BJQ-1（813K）：当某发动机 Ng 值超过最大起飞功率 100.4% 或两台发动机的扭矩和超过 102.5% 时，输出不连续的警告铃声信号给耳机。当某发动机发生故障，两发间扭矩差 >25% 时，锣声产生装置输出连续的警告铃声信号给耳机，用以警告驾驶员操作。同时，极限灯亮。

检查要点：外观、安装、插头、电缆。

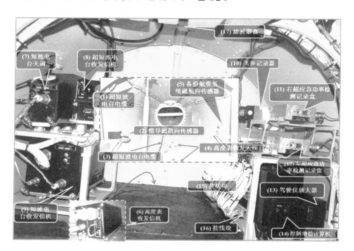

行李舱后部

（1）（3）超短波电台电缆：连接电缆。

检查要点：外观、安装、电缆。

（2）惯导磁航向传感器 HJG-1A（42S）：用于测量载体相对地磁北极的航向，用以指示载体相对地磁北极的偏角，与 HJG-1A 捷联惯导系统配套，为捷联惯导系统提供相对地磁北极偏角的信息。

检查要点：外观、安装、插头、电缆。

（4）高度表收发天线（13S/14S）：发射和接收频率调制信号，测量直升机的高度。

检查要点：外观、安装、插头、电缆。

（5）备份航姿系统磁航向传感器（113F）：测量磁航向参数提供备份航向。

检查要点：外观、安装、插头、电缆。

（6）高度表收发信机 GG0.6-1F（11S）：实时测量直升机相对

地面的真实高度。

检查要点：外观、安装、插头、电缆。

（7）短波电台天调（11R）：实现发射机与天线间的阻抗匹配，使天线在任何频率上有最大的辐射功率。

检查要点：外观、安装、插头、电缆。

（8）超短波电台收发信机（50R）：超短波调频、调幅、抗干扰通信电台的收发信机，用于空–空、空–地间话音通信联络。

检查要点：外观、安装、插头、电缆。

（9）短波电台收发信机（10R）：短波波段接收与发射信号处理。

检查要点：外观、安装、插头、电缆。

（10）飞参记录器：飞行参数防毁记录器。

检查要点：外观、安装、插头、电缆。

（11）（12）右、左超应急功率检测记录盒（114E、113E）：当发动机超应急功率达到5s以上时，超应急检测记录盒上的警告旗由黑色翻转为红色。该发闪光继电器也同时接通，相应的超应急功率警告灯闪亮。盒内还有一个寿命计时器。

检查要点：外观、安装、插头、电缆。

（13）驾驶仪放大器FKJ–10（19C）：一方面根据计算机的指令信号控制俯仰、横滚双马达舵机和偏航配平舵机，另一方面，给计算机提供所需各种信号及监控系统回路。

检查要点：外观、安装、插头、电缆。

（14）控制增稳计算机SJ–2C（16C）：一方面，它处理来自地平仪、捷联惯导、大气数据组件等传感器的信号。另一方面，它可以和耦合操纵台联合，并将处理过的信号送给执行机构，以实现直升机的姿态稳定和自动导航等功能。

检查要点：外观、安装、插头、电缆。

（15）（16）接线块（9ΔA1、9ΔB1）：连接模块。

检查要点：外观、安装、电缆。

（17）滤波器盒ELB–1（650C）：控制增稳系统部件，用于滤除航向操纵舵机的位置反馈信号及±15V电源的高频干扰。

检查要点：外观、安装、插头、电缆。

7 α 告警灯板

TEST 试验	GEN.1 发动机1与汇流条断开（不工作）	BUS.CPL 汇流条相连	GEN.2 发电机2号汇流条断开（不工作）	ROT.BK 旋翼刹车（已处于刹车位）	ENG.1 左发动机滑油压力低（<1.3bar，正常0.8~3bar）	MGB.P 主减速器滑油低（<1.3bar，正常1.8~5bar）	ENG.2 右发动机滑油压力低（<1.3bar，正常0.8~3bar）	CSAS 控制增稳断开
	BAT.SW.1 蓄电池与汇流条PP9断开	FUEL 左/右燃油滤堵塞或燃油压力低于0.2bar	BAT.SW2 蓄电池与汇流条PP8断开	DOORS 舱门未关好	HYD.1 液压系统压力1 低（<10bar，正常60±8bar）	AUX.HYD 辅助液压系统过压（>30bar，正常140bar）	HYD.2 液压系统压力2 低（<10bar，正常60~8bar）	C.P.L 耦合器故障
	INV.1 交流电源系统1故障	SHED.BUS 卸载汇流条（断电）	INV.2 交流电源系统2故障	STEP 脚蹬板在打开位置	OIL.TEMP 主减或发动机滑油温度高（主>132.5℃，发>125℃）	SERVO 主伺服卡滞	BAT.TEMP 电瓶高温（>71±2℃）	
LAND.LT 着陆灯燃亮	GSS.HEAT 速度矢量传感器加温系统故障（未加温）	MAIN.PMP 主减主滑油泵压力低（<0.6bar）	PITOT 空速管加温系统故障（未加温）	O/HEAT 行李舱P2管路过热（>110℃）	CARGO.F 行李舱失火警（>160℃）	L/G.PUMP 起落架应急放系统过压（泵工作）（>95bar/125bar）	FUEL.Q 燃油面低（<18L）	GOV 发动机降级故障（电调轻微故障）
DIM 暗	EXT.L.H 发动机左灭火瓶空	AUX.PMP 辅助滑油泵压力低（<0.6bar）	EXT.R.H 发动机右灭火瓶空	HORN 音响开关断（旋翼转速170~335，380）		HYD.LEVEL 右液压油箱低油位（右<2L尾伺服右切断）		TNG 培训单发训练
	CHIP.1 左发动机屑片探测	O/SPEED.1 左发自由涡轮超转（超转系统故障N+25%灭）	CHIP.2 右发动机屑片探测	O/SPEED.2 右发自由涡轮2超转（超转系统故障N+25%灭）	GOV.1 左发故障	DIFF.NG 功率损失ΔNg>6% 单发故障	GOV.2 右发故障	RPM.365 旋翼超转（>365r/min）

备注：文字带下画线的警告灯为Z9-WZ型直升机所有。

α 组设备安装位置表

设备代号	设备名称	安装位置
1α	主配电盒	无线电设备舱下
2α	断路器板	操纵台左侧
3α	断路器板	操纵台右前侧
4α	断路器板	操纵台右后侧
5α	断路器板	操纵台右侧中部
6α	电源控制板	仪表板下左
7α	告警信号灯板	仪表板上部中间
8α	（全部）仪表板	
9α	组合仪表	新机由综显代替
10α	电压选择开关	6α 上
11α	主告警控制盒	无线电舱
12α	上部电气控制板	座舱顶棚
13α	副驾驶周期变距杆手柄	
14α	正驾驶周期变距杆手柄	
15α	副驾驶总桨距杆手柄	
16α	驾驶员总桨距杆手柄	
17α	a、b 为左右主告警灯	仪表板上
18α	a、b 为左右极限灯	仪表板上
19α	分离插头	行李舱左上
20α	分离插头	X5630 框
21α	左电气设备安装板	行李舱
22α	右电气设备安装板	行李舱
23α	继电器盒	X1560 框左边纵梁
24α	继电器盒	X1560 框右边纵梁
30α	接线盒	X1605 框左侧
31α	接线盒	23α 下面
35α	分离插头	垂直安定面下分离面

第二章　通电检查

通电检查规定

给直升机连接地面电源和通电时应遵守以下规定：

（1）连接地面电源或通电前，必须得到机械师的同意；

（2）接通电源前，应确实判明各配电板已盖好，用电设备的各个电门均应在规定位置，检查蓄电池电压不低于 24V；

（3）机上总电门在断开位置时才能连接地面电源，断开机上总电门后才允许拔下地面电源插头，断开机上总电门前应得到正在通电检查人员的同意；

（4）在下列情况下禁止给直升机连接地面电源或在直升机上通电：

①加添或放出燃油时；

②当电源系统有故障，或检查、拆装电源系统的机件时；

③用汽油擦洗直升机或发动机时；

④拆装燃油系统的导管、附件或燃油系统漏油时；

⑤电压不符合规定值时；

⑥附近上空有强烈雷电时；

⑦可能引起电路短路时；

⑧正在拆卸用电设备时。

一、准备工作

拉起地平仪扶正拉杆（约束 30s，待陀螺转速稳定后松开），同时接通蓄电池 / 地面电源开关 1、2 并按灭主告警灯。此时 7α 板上应有以上告警灯燃亮。

二、电压检查

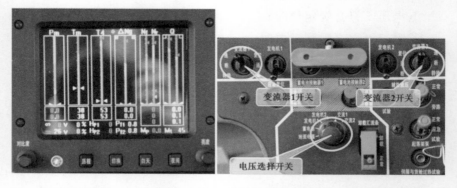

接通变流器 1、2 控制开关。待发动机参数显示器屏幕变亮后，

转动电压选择开关，分别检查地面电源、蓄电池、变流器 1、2 的电压是否符合要求。

三、综合显示系统初始化

（1）将多功能键盘（MFK）综显控制系统开关（IDCS）拨到"STB"位置，待键盘面板上"MFK OK"字母不再闪烁后，拨到"ON"位置，多功能彩色显示器（MFCD）显示红色字母"A IN"。

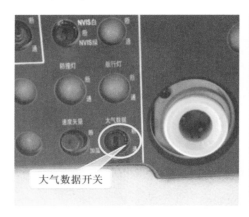

（2）接通大气数据开关，"A"字消失。

四、导航系统对准

精对状态：将惯性导航转换开关"INS"置于"NORM"位置，对准工作方式前 4 分钟显示为"FAST"，4 分钟后显示为"NORM"。

8 分钟后，"ALIGN IN PROGRESS" 的显示切换为 "END OF ALIGN"，表明对准完毕。

此时将 "INS" 开关置于 "NAV" 位置，画面转入 HSD。

快对状态：将惯性导航转换开关 "INS" 置于 "NAV" 位置，对准工作方式显示为 "FAST"，3 分钟左右自动转入 HSD 画面，表明对准完毕。

五、气压高度表检查与火警试验

（1）调节气压高度表使指针归零；

（2）检查左、右火警试验开关（上火警下故障），向上扳动——火警灯亮，主告警灯闪亮；向下扳动——故障灯亮。

六、7α 板检查

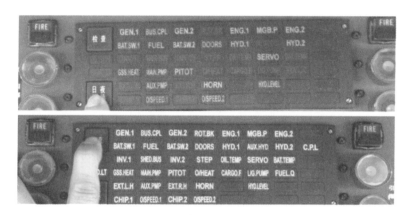

检查 7α 信号板试验（TEST）按钮，8 个灯不亮（左右火警灯、故障灯，左右极限灯、转输泵灯）；按压日夜转换（DIM）按钮，7α 信号板灯变暗；检查左右极限灯，按压变亮。

七、多功能显示器检查

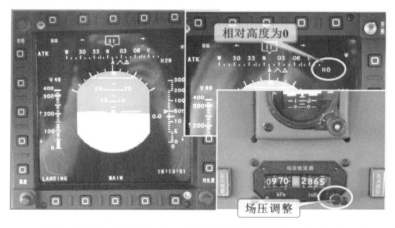

VSD（垂直状态）页面下，调节场压给定器，将直升机相对高度标定为 0，检查场压值与气压高度表场压值应一致。

STATUS（设备状态）页面：1——IDCS（综合显示控制系统），3——HNS（组合导航系统），4——HADS（大气数据系统），6——EPU（发动机参数显示器）；

DATA（数据提取）页面：1——HNS（导航系统数据），2——

HADS（大气数据提取），6——EPU（发动机参数采集数据），9——TANK（燃油数据提取），10——IDCS（综合显示控制系统数据），检查各系统设备状态、参数是否正常。

八、航空时钟与发动机参数显示器检查

检查航空时钟，指针是否归零，走动有无卡滞，时间显示是否准确（1 走 2 停 3 停回零）。

检查发动机参数显示器，分别按压黑夜、白天切换按钮，检查系统状态。

九、6α 控制板检查

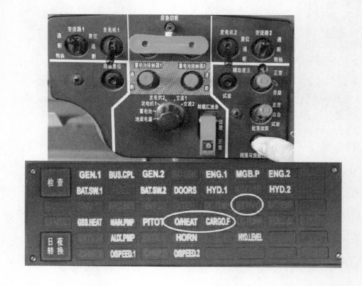

检查 6α 电源控制板，按压伺服、行李舱火警和行李舱过热试验按钮，7α 信号板"伺服"（SERVO）灯灭，"行李舱火警"（O/HEAT）、"行李舱过热"（CARGO/F）灯亮。

十、机外信号灯检查

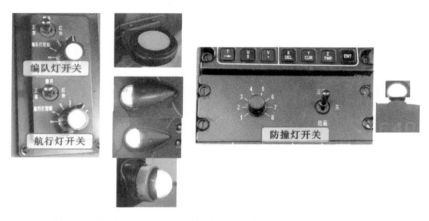

（1）检查防撞灯，接通控制开关，防撞灯闪亮；

（2）检查航行灯，接通控制开关，航行灯亮（左红右绿尾白）；

（3）检查编队灯，接通控制开关，编队灯亮。

十一、燃油操纵台检查

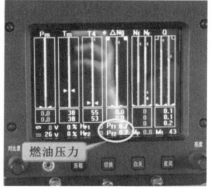

（1）检查燃油操纵台，分别检查 1、2、3、4 号增压泵压力是否正常（0.2~1.2bar），接通 1 或 3 号增压泵开关，左"FUEL.P"灯灭，接通 2 或 4 号增压泵开关，右"FUEL.P"灯灭，同时接通左右

增压泵，7α 板上的"FUEL"警告灯灭；

（2）向左打转输泵开关，左侧转输泵指示灯燃亮，左侧"HI.LEV"灯亮；向右打转输泵开关，右侧转输泵指示灯亮，右侧"HI.LEV"灯亮。

十二、12α 控制板检查

检查 12α 上部控制板，接通速度矢量传感器加温开关，7α 板上的"GSS.HEAT"灯灭；接通空速管加温开关，7α 板上的"PITOT"灯灭（地面加温时间不允许超过两分钟）；接通尾液压隔离切断开关，7α 板上的"HYD.LEVEL"灯灭；接通周期配平开关，按压左右驾驶杆上的释放按钮（DBM）无卡滞，声音清脆；接通音响报警（喇叭）开关，7α 板上的"HORN"灯灭。

十三、总距杆、驾驶杆与脚蹬检查

检查总距杆、驾驶杆、脚蹬；接通周期配平开关，应急电动泵开关拨到试验位置（左系统有压力），抬放总距杆、扳动驾驶杆、蹬脚蹬检查操纵有无卡滞现象；检查结束后，周期杆归中立，脚蹬中立，总距杆放下锁死。

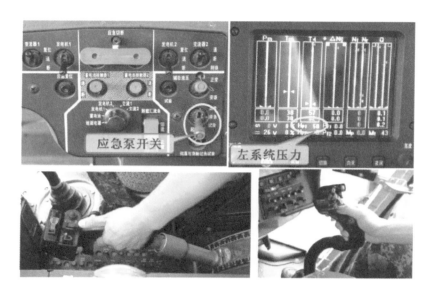

十四、控制增稳系统地面自检

接通条件：

（1）地平仪故障旗消失，1回路三个通道故障黄灯熄灭；

（2）惯性导航完成对准，2 回路三个通道故障黄灯熄灭；

（3）自驾操纵台无指示灯亮；

（4）接通变流器 1、2，打开周期配平开关，应急电动泵开关打到试验位置。

（1）同时接通 1、2 回路按钮，自检开关打到"RUN"位置，自动驾驶仪自检。

（2）待七段数字指示器显示"0"，断开自检开关，待舵面位置指示器全部归中立后，同时断开 1、2 回路按钮，断开应急电动泵、周期配平开关。

地面自检注意事项：

（1）地面通电检查时，操纵系统必须加液压；

（2）地面自检过程中，中途不要打断自检；

（3）舵面位置指示器中立时，方可断开自驾；

（4）先断开自驾，再断开液压；

（5）空中禁止自检。

十五、耦合器检查

接通自驾 1、2 回路，接通耦合器操纵台按钮"CPL"，接通耦合器操纵台按钮"A/S"，此时处于空速保持状态，驾驶杆自动向前运动，当数显空速表数字显示为"74"时，驾驶杆停止运动。

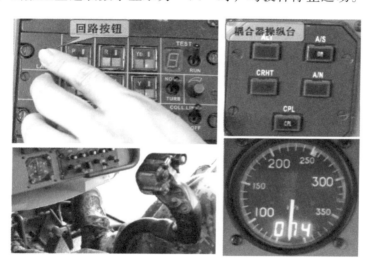

十六、灯光照明系统检查

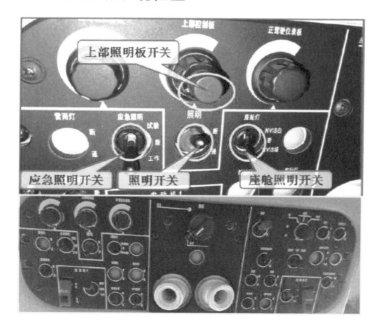

　　将应急照明试验开关置"工作"位，照明开关置"通"位，座舱照明开关分别置"绿"和"白"，将上部照明板开关和仪表板照明开关转到最大位置，检查仪表板、操纵台、4α、5α、12α板照明情况；检查活动灯照明；接通着陆灯开关，7α上的"LAND.LT"灯亮，着陆灯燃亮，检查着陆灯收放情况。（跨昼夜飞行时检查。）

十七、通信系统检查

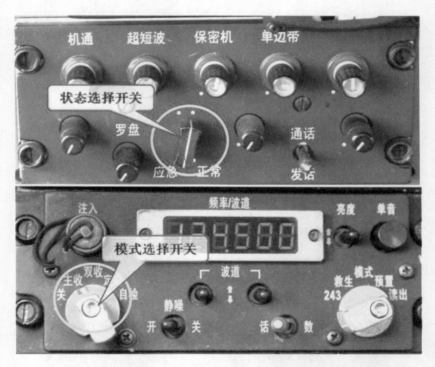

　　将机通控制盒上的状态选择开关拨到"应急"位置，将无线电台控制盒面板上的模式选择开关由"关"位旋转至"主收"位置，电台开始自检。频率/波道显示窗应显示自检结果，自检通过后应显示为"PASS"，随后频率/波道显示窗应显示波道号。

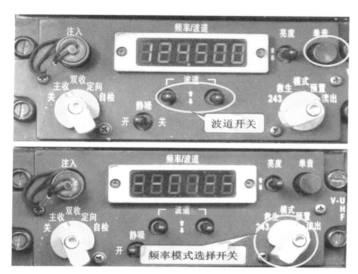

自检通过后，当前电台工作使用的波道号将显示在频率/波道显示窗上，耳机中应有噪声或音频信号。随后按照任务要求调整当前工作波道，然后将频率模式选择开关置于"读出"位置，调整该波道工作频率并按"单音"键确认。最后将频率模式选择开关置于"预置"位置。此时耳机中应有噪声或音频信号。

将静噪开关置于"开"位置，此时耳机中应无噪声。如当前无线

电台正在接收音频信号，则信号间隙应无噪声，随后将静噪开关置于"关"位置。

　　将模式选择开关置于"双收"位置，此时主接收机在预置的模式下工作，救生接收机在预置频率对应的救生频率及模式下工作，耳机中有噪声或话音。

　　注意：抗干扰方式下，救生接收机不工作。

　　将控制盒工作模式选择开关置于"主收"位置，频率模式选择开关置于"救生"位置，此时主接收机在预置频率对应的救生频率上工作，救生机不工作，耳机中可听到噪声。

　　将工作模式选择开关置于"主收"位置，频率模式选择开关置于"预置"位置并按下"单音"按钮，此时发射机工作并可以从耳机中听到单音信号。

　　注意：抗干扰方式和30MHz~88MHz频段无单音功能。

　　将控制盒工作模式选择开关置于"关"位置，显示窗应不亮；频率模式选择开关置于"243"位置，显示窗应显示"243.000"。此时

主接收机、发射机全部在 243.000MHz 频率上以"调幅"方式工作。

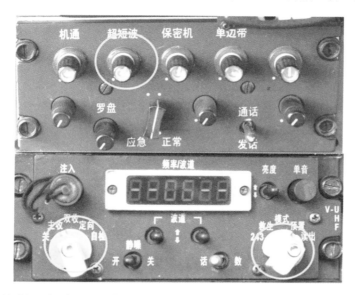

将控制盒频率选择模式开关置于"预置"位置，工作模式选择开关置于"主收"位置，电台开始自检。自检结束后，按下 JT-9 机内通话器上的"超短波"收发通道，按压驾驶杆上的 PTT 键至 2 挡，然后发话至工作在制定频率的电台，此时耳机中应有自听声音。随后松开 PTT 键并接收对方的回答信号。

检查完毕后，将更改过的波道频率恢复至初始频率，再将工作波道调整至开机前的波道上（01波道），然后关闭电台。机通控制盒状态开关置于"正常"位置，复位其他开关和电门。

十八、应急地平仪检查

·分别接通4α上的断路器A4、5α上的断路器G2，检查应急地平仪工作情况，常温下（5℃~20℃），3分钟内，应急地平仪告警旗应收起。

·拉出应急地平仪调整旋钮，使应急地平仪移至零位，此时，告警旗应出现；松开调整旋钮，告警旗应消失，应急地平仪仍在零位。

·断开断路器，告警旗重新出现。

注意：地平仪通电 3min 后检查：

·地平线应在飞机标志符中心圆之内。

·俯仰指引杆和倾斜指引杆应准确地交叉在飞机标志符中心圆之内。

·俯仰指引杆与飞机标志符的平行度应小于±3mm，且倾斜指引杆相对于倾斜刻度的零标记小于±3mm。

第三章 专项检查

一、KZW-3 控制增稳系统的检查

1. 所需设备

地面电源设备或等效设备。

2. 准备工序

· 打开接近控制增稳系统各部件的检查口盖。

· 连接地面电源设备。

3. 断开电源检查

· 对于控制增稳系统的每个部件必须检查：

一般状况、有无变形、清洁度、安全可靠性、腐蚀情况、标牌及密封的情况（对于密封的部件而言）；电气接插件的状况和连接情况，检查其标记的可见性。

除了上述检查以外，还要检查部件的支架。对于下列附件进行下述检查：

· 电路中的断路器：检查规格和连接情况。

· 位置指示器：电气连接状况、指针、刻度情况。

· 控制增稳系统和耦合器的"主警告灯"：电气连接情况、判读性。

· 耦合器指示灯：机械状况、刻字的情况、电气连接情况。

· 驾驶员和副驾驶员周期变距操纵杆和总桨距操纵杆上的控制按钮：机械状况。

· 大气数据计算器：螺母、管路（一般状况、清洁情况、腐蚀情况、锁紧情况）。

·配平舵机（并联舵机）：输入连杆（一般状况、清洁情况、腐蚀情况、有无变形、与操纵连杆和摇臂的连接情况、叉耳端头和开口销的状况及清洁情况）。

·电动舵机（串联舵机）：轴和曲柄上舵机的连接及开口销情况；半连杆（安全可靠性、清洁情况、腐蚀情况、一般情况、连拉螺栓的锁紧情况）。

·开关拉杆：机械情况，电气连接。

·导线、接线排、接地接线排、接地点、电气接插件、护套、屏蔽、隔板和地板接插件、搭铁线与结构间的连接、护套内的导线、卡子中的护套、与结构连接的卡子。

·一般情况、腐蚀情况、安全可靠性。

4. 直升机通电检查

接通电源并进行下列检查：

·控制增稳系统的耦合操纵台：机构状态、刻字和刻线状态、检查照明电位计的工作；检查指示灯的灯光、刻字和刻线及警告灯的正确性和规则变化。

·电流计：检查仪表板照明电位计的工作；检查电流计的照明是否正确，并是否规则的变化。

·控制增稳系统主警告灯：按压仪表板上的"TEST（试验）"按钮，检查警告灯是否正确燃亮。

注：在操纵控制增稳操纵台上自动配平开关时，7α 上"CSAS"灯闪亮属正常。

5. 最后工序

·断开电源，断开地面电源装置。

·关闭检查口盖。

二、KZW-3 控制增稳系统的自检程序

1. 所需设备

地面电源装置或等效设备。

注：控制增稳系统用单一方式长期工作需连接上地面液压动力源或等效设备。

2. 准备工序

连接地面电源，进而向直升机系统提供 28V 直流电源及 115V/400Hz、26V/400Hz 的交流电源。

舵机处于中立位置，陀螺地平仪和组合导航系统的惯性测量部件正常工作。

3. 试验：回路 1，回路 2

· 接通控制增稳系统的所有通道和功能（控制盒上的所有开关处于上面位置）。

· 将 "TEST（试验）" 开关自凹槽脱开，并置于下部 "RUN（运行）" 位置。

· 在整个试验（约 30s）中，显示窗口的小数点一直在闪烁，六个黄色灯（P1、P2、R1、R2、Y、YD）和三个自动配平指示灯（AUTO TRIM）随着检验程序的进行而闪亮，继六个通道和三个自动配平指示灯第二次循环后，显示窗口的小数点消失，即指示该试验结束。

· 如果控制增稳系统被验证正确，则试验结束时指示窗口应显示 "0"。

· 如果有一个或多个功能有故障，则显示出对应的数码。记下这些数码，然后参考制造厂家的手册便能判别故障。

· 将 "TEST（试验）" 开关扳到上部位置，并卡入凹槽内。

· 指示灯熄灭。

· 等到试验结束为止。

4. 断开回路 1

· 将试验开关置于 "RUN（运行）"。

· 在单一方式试验程序期间 "TRIM（配平）" 和 "COLL LINK（总距）" 信号灯应至少燃亮一次。

· 当该程序完成以后，将试验开关置于 "TEST（试验）" 位置：俯仰和横滚电流计应像先前那样摆动；但摆幅只有原先摆幅的一半。

注：不必考虑显示的数字。

5. 断开回路 2，接通回路 1

· 关闭 "COLL LINK（总距预控）" 的工作，然后把 "TEST（试验）" 开关放到 "RUN（运行）" 位置。

·检查该程序结尾出现的号码1~7。

·把开关放到"试验（TEST）"位置，按通"总距预控（COLL LINK）"的工作位置。

·重新接通回路2。

6. 模拟飞行条件

·使9G和10G接触偶的D极和E极连接。

·把"TEST（试验）"开关放到"RUN（运行）"位置，该试验程序不应开始。

·把"TEST（试验）"开关放到"TEST（断开）"位置。

7. 最后工序

·拆掉9G和10G之间的短路线。

·断开电源，断开地面电源装置。

·降低液压系统的压力，将地面液压动力装置断开。

三、交流电源系统试验

1. 所需设备

无。

2. 准备工序

·接通直流电源系统。

·断开"变流器1、2（INVERT1和INVERT2）"控制开关。

·信号板上的"变流器1（INV.1）"和"变流器2（INV.2）"指示灯燃亮。

3. 试验程序

·接通"变流器1、2（INVERT1和INVERT2）"控制开关。

·信号板上的两个指示灯应熄灭。

·通过选择开关10α（置于INVERT1和INVERT2）检查电压表电压读数是否为115V。

·将"变流器1（INVERT1）"开关置于"断开（OFF）"位置，再置于"转换（TRF）"位置，26V和115V系统应保持供电，恢复到正常状态。

·将"变流器2（INVERT2）"开关置于"断开（OFF）"，然后再置于"转换（TRF）"位置，检查 26V 和 115V 系统应保持供电，恢复到正常状态。

4. 最后工序

· 断开"变流器 1、2（INVERT1 和 INVERT2）"开关。

· 断开直流电源。

四、直流电源系统试验

1. 所需设备

28V 直流地面电源装置。

WQ1 型 0~40V 直流可调稳压电源。

2. 发电机系统试验

（1）准备工序。

· 插上地面电源装置插头。

· 起动发动机。

· 置"蓄电池接触器 49P（BAT.RLY1）、50P（BAT.RLY2）"于"接通(ON)"位置，则"蓄电池接触器 1（BAT.SW1）""蓄电池接触器 2（BAT.SW2）""发电机 1（GEN.1）""发电机 2（GEN.2）"和"汇流条相连（BUS.CPL）"指示灯均燃亮。

（2）程序。

· 拔下地面电源插头。

注：当发电机转速达到额定值时，才能拔下地面电源插头。

· "蓄电池接触器 1、2（BAT.SW1、BAT.SW2）"和"汇流条相连（BUS.CPL）"指示灯熄灭

发电机 1（左）。

· 置"发电机 1(GEN.1)"控制开关于"复位（RST）"位置，然后于"接通(ON)"位置。"发电机 1（GEN.1）"指示灯熄灭，"汇流条相连（BUS.CPL）"指示灯燃亮。

· 置电压表选择开关 10α 于"发电机 1（GEN.1）"位置，电压表读数为 28.5V 或 27.5V，这取决于调压保护器的调整。

注：如果电压表读数不正确，则调整调压保护器。

· 置"蓄电池接触器 1（BAT.RLY1）"开关于"接通（ON）"位置。

而"蓄电池接触器2（BAT.RLY2）"开关于"断开（OFF）"位置。

·"蓄电池接触器1（BAT.SW1）"指示灯熄灭，而"蓄电池接触器2（BAT.SW2）"指示灯燃亮，"汇流条相连（BUS.CPL）"指示灯熄灭。

·1s后，置"蓄电池接触器2（BAT.RLY2）"开关于"接通（ON）"位置。检查"蓄电池接触器2（BAT.SW2）"指示灯仍保持燃亮，而"汇流条相连（BUS.CPL）"指示灯仍保持熄灭。

·同时置"蓄电池接触器1、2（BAT.RLY1和BAT.RLY2）"二开关于"接通（ON）"位置，相应的指示灯均熄灭，而"汇流条相连(BUS.CPL)"指示灯燃亮。

发电机2（右）：

·置"发电机1（GEN.1）"开关于"断开（OFF）"位置，"发电机1（GEN.1）"指示灯燃亮。

·置"发电机2（GEN.2）"开关于"复位（RST）"位置，然后于"接通（ON）"位置，"发电机2（GEN.2）"指示灯熄灭，"汇流条相连（BUS.CPL）"指示灯燃亮。

·进行左（1号）发电机规定的所有试验步骤。

卸载系统试验：

·接通两台发电机。

·用一块电压表，检查发电机1和发电机2的电压（汇流条PP8和PP9均有电）。

·置卸载汇流条开关于卸载（SHED）位置，则7α板上的卸载汇流条指示灯燃亮，4α板上的PP10F汇流条(D10~D15)、PP13F汇流条(A6~A15)、5α板上的PP15F汇流条(E1~E10)和2α板PP12F汇流条(K1~K9)不再有电。

注：这些汇流条实际卸载还可以靠检查连接到这些板的断路器上的设备不再有电来证实。

发动机停车：

·置蓄电池接触器1、2的开关于断开（OFF），所有的指示灯熄灭；

·置卸载汇流条开关于正常（NORM），并合上保险盖。

3. 蓄电池和地面电源插座系统试验

（1）准备工序。

·发动机停车状态。

·置"蓄电池接触器1、2（BAT.RLY1 和 BAT.RLY2）"和"发电机1、2（GEN.1 和 GEN.2）"开关于"断开（OFF）"位置。

·检查信号板上所有的指示灯应熄灭。

·插上地面电源装置插头。

警告：地面电源插头无论是插上或拔下总是带电的。

（2）程序。

地面电源系统试验：

·置"蓄电池接触器1、2（BAT.RLY1 和 BAT.RLY2）"开关于"接通（ON）"位置。

·检查"发电机1、2（GEN.1 和 GEN.2）"指示灯和"汇流条相连（BUS.CPL）"指示灯及"蓄电池接触器1、2（BAT.SW1 和 BAT.SW2）"指示灯均燃亮。

·置"蓄电池接触器1（BAT.RLY1）"开关于"断开（OFF）"，而"蓄电池接触器2（BAT.RLY2）"开关于接通（ON），并且反过来做一次。

·检查"发电机1（GEN.1）"和"发电机2（GEN.2）"指示灯和"汇流条相连（BUS.CPL）"指示灯及"蓄电池接触器1、2（BAT.SW1 和 BAT.SW2）"指示灯仍保持燃亮。

·将选择开关10α置于"地面电源（EXT）"位置，检查发参显示器应指示地面电源装置的电压。

·置两个"蓄电池接触器1、2（BAT.RLY1 和 BAT.RLY2）"开关于"断开（OFF）"位置。

·拔下地面电源插头。

蓄电池系统试验。

·同时将"蓄电池接触器1、2（BAT.RLY1 和 BAT.RLY2）"开关置于"接通（ON）"位置。

·将选择开关10α置于"蓄电池（BAT）"位置，检查发参显示器应显示蓄电池的电压。

·检查"蓄电池接触器1、2（BAT.SW1 和 BAT.SW2）"指示灯和"汇流条相连（BUS.CPL）"指示灯不应燃亮。

注：如果两个开关不同时动作，后动作的一个蓄电池接触器不能接通。因为短路保护系统在相应汇流条上探测出一个低电压故障，从而阻碍了第二个接触器接通。汇流条相连接触器也不能接通。

特别注意：试验期间，检查蓄电池温度警告灯不应燃亮。

4. 过压保护和复位电路试验

（1）过压保护。

·起动发动机（此试验与发电机电路试验一起进行更好）。

·在空载状态下，按压调压保护器（28P 或 29P）的散热片中心处的小红色按钮。这就接通了造成过压状态的一个电路系统。1 秒后，发电机输出电压大约升至 32.5V，然后过压保护电路将调压保护器激磁电路断开，并断开发电机激磁。发电机输出电压跌落至 1V 以下，且"发电机 1（GEN.1）"或"发电机 2（GEN.2）"指示灯燃亮。

·用"发电机 2"（GEN.2）或"发电机 1"（GEN.1）开关（6α 板上）复位调压保护器，则系统电压恢复到 28.5V，且发电机反流保护器接通。

（2）应急复位。

·置"蓄电池接触器 1、2（BAT.RLY1 和 BAT.RLY2）"开关于"断开（OFF）"位置。

·置"发电机的 1、2（GEN.1 和 GEN.2）"开关于"断开（OFF）"位置。

·再把"发电机 1、2（GEN.1 和 GEN.2）"开关置于"接通 (ON)"位置。

·按压"应急复位（EMERG.RESET）"按钮。

·在外接电压表上检查每台发电机电压表读数应正常地增加。

·置"蓄电池接触器 1、2（BAR.RLY1 和 BAT.RLY2）"开关于"接通（ON）"位置。

·检查"发电机 1、2（GEN.1 和 GEN.2）""汇流条相连（BUS.CPL）"和"蓄电池接触器 1、2（BAT.SW1 和 BAT.SW2）"指示灯熄灭。

5. 地面电源插座保护试验

·拔下蓄电池插头。

·连接一个 0 到 40V 直流可变电源到直升机地面电源插座接线柱上：电源正接线柱接到地面电源插座正接线柱和小接线柱上；电源负接线柱接到地面电源插座负接线柱上。

·调整可变电源电压约为直流 28V。

·置"蓄电池接触器 1（BAT.RLY1）"开关于"接通（ON）"位置。

·接通断路器 11P，直升机供电，外接电压表指示地面电源电压。

·慢慢地增加可变电源电压。

·当电压为 33±1V，直升机不再供电，机上发参显示器黑屏。

·使可变电源电压返回到 28.5V，直升机一直未供电。

·断开断路器 11P。

·接通断路器 11P，直升机重新供电，机上发参显示器指示可变电源电压。

·断开断路器 11P，直升机不再供电，机上发参显示器黑屏。

·置"蓄电池接触器 1（BAT.RLY1）"开关于"断开（OFF）"位置。

·使可变电源电压为零，并使其与地面电源插座断开。

·重新接上蓄电池插头。

五、直流电源系统发电机电压调整和均衡检查

1. 所需设备

C31-VA 型 1.5 级直流电压表。

注：此操作是在首次地面运转调整和每次更换调压保护器时进行。

2. 准备工序

·直升机正常安装好，蓄电池应在指定位置上并已充好电。

·打开行李舱门。

·打开左、右电气设备安装板。

·置"蓄电池接触器 1、2（BAT.RLY1 和 BAT.RLY2）"开关和"发电机 1、2（GEN.1 和 GEN.2）"开关于"接通 (ON)"。

·起动发动机且增加转速到正常额定功率。发电机必须在这种额定状态下投入电网（发电机信号灯熄灭）。

注：当外界气温达到或超过 35℃时，推荐直升机电压调至 27.5±0.5V。否则，调至 28.5±0.5V。

3. 程序（下图）

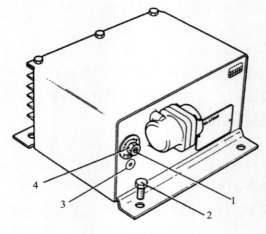

发电机调压器

（1）空载状态时发电机电压的检查和调整。

·置"发电机 1、2（GEN.1 和 GEN.2）"开关于"断开（OFF）"位置。"发电机 1、2（GEN.1、GEN.2）"指示灯燃亮。

·把电压表先接到左边，然后再接到右边：正线接至所涉及的调压器的试验插座接线柱（3）上，而负线接到直升机结构上（调压器紧固螺钉（2））。

·记下每台发电机的电压。

·如果必要，调整电压到 28.5V。松开锁紧螺母（4），拧动调整螺钉（1），以调整电压（按要求 28.5±0.5V 或 27.5±0.5V）。拧紧锁紧螺母，检查电压应无变化，反之重新调整。

·断开电压表。

·座舱内，置"发电机 1、2（GEN.1 和 GEN.2）"开关于"接通(ON)"位置，"发电机 1、2（GEN.1 和 GEN.2）"指示灯熄灭。

·关上左、右电气设备安装板。

·关上行李舱门。

（2）均衡检查和调整。

置 6α 板上的电压表选择开关 10α 于"发电机 1（GEN.1）"位置。

·置"蓄电池接触器 1（BAT.RLY1）"开关于"断开（OFF）"位

置,"蓄电池接触器 1(BAT.SW1)"指示灯燃亮。

·断开"发电机 1(GEN.1)"开关,"发电机 1(GEN.1)"指示灯燃亮。

·检查发电机 1 电压值应与上述调节电压值相符合。

·置"发电机 1(GEN.1)"开关于"接通(ON)"位置,"发电机 1(GEN.1)"指示灯熄灭。

·置"发电机 2(GEN.2)"开关于"断开(OFF)"位置,"发电机 2(GEN.2)"指示灯燃亮。

·检查发电机 1 的输出电流。

·置"发电机 2(GEN.2)"开关于"接通(ON)"位置,"发电机 2(GEN.2)"指示灯熄灭。

·同时置"蓄电池接触器 1、2(BAT.RLY1 和 BAT.RLY2)"开关于"接通(ON)"位置。

·"蓄电池接触器 1、2(BAT.SW1 和 BAT.SW2)"指示灯熄灭。置电压表选择开关 10 α 于"发电机 2(GEN.2)"位置。

·置"蓄电池接触器 2(BAT.RLY2)"于"断开(OFF)"位置,"蓄电池接触器 2(BAT.SW2)"的指示灯燃亮。

·置"发电机 2(GEN.2)"开关于"断开(OFF)"位置,"发电机 2(GEN.2)"指示灯燃亮。

·检查电压表上的发电机 2 的电压值应与上述调节电压值相符合。

·再置"发电机 2(GEN.2)"开关于"接通(ON)"位置,"发电机 2(GEN.2)"指示灯熄灭。

·置"发电机 1(GEN.1)"开关于"断开(OFF)"位置,"发电机 1(GEN.1)"指示灯燃亮。

·检查发电机 2 的输出电流。这个值实际上应与发电机 1 的值相等。

·置"发电机 1(GEN.1)"开关于"接通(ON)","发电机 1(GEN.1)"指示灯熄灭。

·同时置"蓄电池接触器 1、2(BAT.RLY1 和 BAT.RLY2)"开关于接通(ON)位置。

·"蓄电池接触器 1、2(BAT.SW1 和 BAT.SW2)"指示灯熄灭。

·检查发电机 2 的输出电压和电流,然后将电压表选择开关 10 α

扳到"发电机 1（GEN.1）"位置。

· 检查发电机 1 的输出电压和电流。

· 在正常工作状态下，发电机 1 的电压与发电机 2 的电压相等。且在这种状态下，每台发电机大约输出电流为 40A（即 25%）时，输出电流之差应不超过 10A。否则，重新调节电压。

· 减少承载较大的那台发电机的电压，以便减少这台发电机的负载。

六、蓄电池温度探测器的检验和检查

1. 所需设备

加热装置（建议用电炉）、两个容器（0.5L 和 1L）、温度计（精度 0.5℃）、三用表、导线和连接销。

2. 准备工序

· 拆下蓄电池，并将蓄电池放在工作台上。

· 拆卸蓄电池盖。

· 从蓄电池单格电池上拆下温度探测器（2 个螺母和垫圈）。

· 在 0.5L 容器里加水至容积的 3/4，并放在电炉上。

3. 程序（下图）

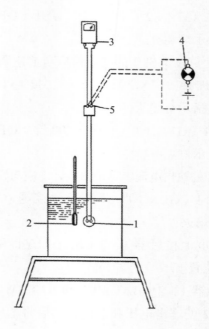

·将温度探测器组件浸入水中，并确保温度探测器（1）不与容器接触。

·将温度计放入水中，并使温度计泡（2）靠近温度探测器。

·将三用表（3）或导通试验器（4）连到温度探测器插座（5）的两个触点上，插座两触点分别接于开关两极上。

·将水慢慢加热，大约水温在60℃时保持10min，以便使所有温度探测器组件达到水温。

·继续慢慢地加热水，并注意当温度探测器闭合（温度升高）时温度计的读数。

·当温度探测器闭合时，三用表（3）指针从"无穷大"摆到"零"或导通试验器灯（4）燃亮。温度探测器正确的升温闭合点为68℃~74℃。

·一旦温度探测器闭合，移开电炉且将0.5L容器放到1L容器中。在大容器里加冷水，直到两容器水面一样高。

·当温度探测器断开时记下温度（降温）。当温度探测器断开时，三用表指针将回到"无穷大"，或导通试验器灯（4）熄灭。温度探测器正确的降温断开点低于升温闭合点温度4℃~10℃。

·如果温度探测器断开、闭合点在公差极限之外，则重复进行上述操作。

4. 最后工序

·从水中取出温度探测器，并彻底干燥。

·将温度探测器固定到蓄电池单格电池上（2个螺母、垫圈）。

·盖好蓄电池盖，并安装蓄电池。

七、直流电源系统短路保护插件功能试验

1. 设备

无。

2. 程序

注：直升机地面工作期间，发动机运转时，不准拆卸任何设备。

（1）右电源系统。

·左、右两台发电机均接入电网，"发电机1（GEN.1）""发电机

2（GEN.2）"和"汇流条相连（BUS.CPL）"指示灯均不燃亮。

· 断开 4α 断路器板上的 C17 断路器。

· 将"发电机 2（GEN.2）"控制开关置于"断开（OFF）"位置，"发电机 2(GEN.2)"指示灯燃亮。

· 重新接通断路器 C17。

· 将"发电机 2（GEN.2）"控制开关置于"复位（RST）"位置。

· 检查发电机应接入电网，"发电机 2（GEN.2）"指示灯熄灭。

（2）左电源系统。

· 实施与上述右电源系统相同的试验，但断路器为 4α 板上的 B17。

注：如果上述试验程序顺利完成，则可认为短路保护插件 40P 和 41P 电路是完全正确的。

八、双供电电路二极管检查试验

1. 所需设备

直流 28V 地面电源装置。

2. 第一次试验

（1）准备工作。

· 在断路器 4α 上断开断路器：

A4、A8、B1、B9、B14、B13、B16、B11、B12、A10、A15。

· 在断路器 5α 上断开断路器：

F1、F2、F3、F4、F5、F6、F7、F8、F9、F10。

· 在断路器 2α 上断开断路器：

M2。

· 进行发动机地面运转。

（2）检查和措施。

· 随着旋翼转速的增加，旋翼处于最大或最小转速时检查音响警报器的声音，如果声音不好，则更换 2α 板内的二极管 28E 或断路器 L3。

· 当开关 38X 和 39X（左右变流器控制开关）置于接通位置时，检查 250VA 交流系统电源应有输出，7α 信号板上的指示灯"变流器（INV.1）"和"变流器（INV.2）"应由燃亮到熄灭，否则检查二极管 46X 和 4α 板上的断路器 D8。

· 断开 12α 板上开关 8F，检查驾驶员用的全压管的温度。然后接通开关 8F，如果温度不升高，则检查 3α 板内的二极管 D6 和 4α 板上的断路器 D6。

· 检查燃油油量表 1Q 的显示情况，如果没有显示或显示不正常，则更换该油量表。

· 接通 3Q 板上转输开关，通过油量表的读数检查燃油的转输情况。如转输情况不好，则检查 2α 板上的二极管 D4 和断路器 L6。

· 按下信号板 7α 上的试验按钮，检查起落架警告灯 3G 应闪亮。如不闪亮，则检查 2α 板上的二极管 56G 和断路器 L9。

· 把开关 8E 置于"火警（FIRE）"位置，检查"火警（FIRE）"警告灯 2E 应燃亮。如果 2E 不亮，则检查 3α 板上的二极管 D3 和 2α 板上断路器 L4。

· 检查控制增稳系统工作是否正常，如不能正常工作，则检查 4α 板上断路器 D10。

· 检查应急地平仪工作是否正常，如不能正常工作，则检查 5α 板上断路器 G2。

· 检查超应急功率系统工作是否正常，如不能正常工作，则检查 5α 板上断路器 G3。

· 检查发参系统工作是否正常，如不能正常工作，则检查 4α 板上的断路器 C5。

· 检查组合导航系统工作是否正常，如不能正常工作，则检查 5α 板上断路器 H6。

· 按下试验按钮 2D，检查 7α 板上的"行李舱火警（CARGO. F）"和"过热（O/HEAT）"警告灯应燃亮。否则，检查 32Δ 上二极管 3W 和 4α 板上断路器 C9。

· 检查风挡雨刷工作是否正常。如不能正常工作，则检查 3α 板上的二极管 D1 和 4α 板上断路器 C16。

· 机上被供电后，主警告灯 (ALARM) 应闪亮，7α 信号板上大部分指示灯应燃亮。如果不亮，则检查 2α 板上的断路器 L8。

· 机上被供电后，接通 12α 板上的"照明"控制开关，仪表板和操纵台上的仪表及断路器板，顶棚控制板等均有照明。如果没有，则检查 4α 板上的断路器 D15。

· 机上被供电后，7α 信号板上警告灯"液压油面（HYD.LEVEL）"应燃亮。如不亮，则检查 4α 板上的断路器 C11 及二极管 D28。

· 断开应急照明蓄电池，将 12α 板上"应急照明"置于"工作"位置，将"座舱照明"及"指令灯"控制开关置于"接通"位置，座舱顶灯及指令灯应燃亮，如不亮，则检查 4α 板上的断路器 C15 及 126L 支架上的二极管 D1、D2、D3、D4。

· 停止发动机地面运转，由地面电源插座为机上供电。

· 按下靠近地面电源插座的按钮 20G，检查灯 24G 燃亮情况：

Δ 如灯 24G 不亮：

拔下地面电源插头，接通 5α 板上的断路器 F8。

插上地面电源插头，按下按钮 20G；若灯 24G 亮，则更换 19G 装置中的二极管 D11 或 1α 中的 13P。

Δ 若灯 24G 仍然不亮：

拔下地面电源插头，断开断路器 F8，并接通 5α 板上的断路器 F9。

插上地面电源插头，按下按钮 20G，若灯 24G 亮，则更换 2α 板上的二极管 D9 或 5α 板上的断路器 G9。

（3）最后工序。

· 接通断开的所有断路器。

· 将操作过的所有开关恢复到原来位置。

· 拔下地面电源插头。

3. 第二次试验

（1）准备工序。

· 断开主配电盒 1α 中的断路器 13P。

· 断开断路器板 2α 上的断路器 L1、L2、L3、L4、L6、L7、L8 和 L9。

· 断开断路器板 4α 上的断路器 C9、C16、C14、D3、D4、D6、D8、D15、C11、C15、C5、D10。

· 断开断路器板 5α 上的断路器 G9、G3、H6、G2。

· 进行发动机地面运转。

· 检查控制增稳系统工作是否正常，如不能正常工作，则检查 4α 板上断路器 A10。

·检查应急地平仪工作是否正常，如不能正常工作，则检查 4α 板上断路器 A4。

·检查超应急功率系统是否正常，如不能正常工作，则检查 5α 板上断路器 F6。

·检查发参系统工作是否正常，如不能正常工作，则检查 2α 板上断路器 M2。

·检查组合导航系统工作是否正常，如不能正常工作，则检查 5α 板上断路器 E9。

（2）检查和措施。

·随着旋翼转速的增加，旋翼处于最大或最小转速时检查音响警报器的声音，如声音不好，则检查 4α 板上的二极管 27E 和 4α 板上的断路器 B13。

·当开关 38X 和 39X 置于接通位置时，用机上电压表检查 $2\times1\,000VA$ 交流电源系统有输出电压。7α 信号板上指示灯"变流器 1（INV.1）"和"变流器 2（INV.2）"应由燃亮到熄灭，否则检查 4α 板上的二极管 47X 和断路器 A8。

·断开 12α 板上的开关 8F，检查驾驶员侧的全压管的温度。然后接通开关 8F。此时全压管温度应升高，若不升高，则检查 3α 板上的二极管 D9 和 4α 板上的断路器 B1。

·检查燃油油量表显示情况，如果没有显示或显示不正常，则更换该油量表。

·接通 3Q 板上的转输开关，通过油量表的读数检查燃油的转输情况。如果转输情况不好，则检查 2α 板上的二极管 D1 和 5α 板上的断路器 F3。

·按下信号板 7α 上的试验按钮。检查起落架警告灯 3G 应闪亮。若不闪亮，则检查 5α 板上的二极管 57G 和断路器 F10。

·把开关 8E 置于"火警（FIRE）"位置，检查"火警（FIRE）"警告灯 2E 应燃亮。若 2E 不亮，则检查 2α 板上的二极管 D2 和 5α 板上断路器 F5。

·把开关 9E 置于"火警（FIRE）"位置，检查"火警（FIRE）"警告灯 1E 应亮，若 1E 不亮，则检查 3α 板上二极管 D2 和 5α 板上断路器 F7。

· 按下试验按钮 2D，检查 7α 板上的"行李舱火警（CARGO. F）"和"过热 (O/HEAT)"警告灯应燃亮。否则，检查 32Δb 上二极管 2W 和 4α 板上断路器 B9。

· 检查风挡雨刷工作是否正常，如不能正常工作，则检查 3α 板上的二极管 D4 和 4α 板上的断路器 B16。

· 检查航行灯是否燃亮，如灯不亮，则检查板 3α 板内 4Δ 上的二极管 D7 和 4α 板上的断路器 B14。

· 机上被供电后，主警告灯 (ALARM) 应闪亮，7α 信号板上大部分指示灯应燃亮，如果灯不亮，则检查 4α 板上的断路器 B12。

· 机上被供电后，接通 12α 板上的"照明"控制开关，仪表板和操纵台上的仪表及断路器板、顶棚控制板等均有照明，如果没有，则检查 4α 板上的断路器 A15。

· 机上被供电后，7α 信号板上的警告灯"液压油面 (HYD.LEVEL)"应燃亮。如果不亮，则检查 4α 板上的断路器 B11 及二极管 29D。

· 接上照明应急蓄电池，将 12α 板上"应急照明"置于"工作"位置，将"座舱照明"及"指令灯"控制开关置于"接通"位置，座舱顶灯及指令灯应燃亮。如不亮，则检查照明应急蓄电池工作是否良好。

· 停止发动机地面运转。

· 按下靠近地面电源插座的按钮 20G，检查灯 24G 燃亮情况：

Δ 如灯 24G 不亮：

则拔下地面电源插头，接通 5α 上断路器 G9(由蓄电池供电)。

插上地面电源插头，按下按钮 20G，若灯 24G 亮，则更换 2α 板内二极管 D6 或 5α 板上断路器 F9。

Δ 若灯 24G 不亮：

拔下地面电源插头，断开 5α 板上的断路器 G9，并接通主配电盒 13P（由蓄电池供电）。

插上地面电源插头，按下按钮 20G，若灯 24G 亮，则更换 19G 装置中二极管 D10 或 5α 板上断路器 F8。

（3）最后工序。

· 拔下地面电源插头。

· 接通断开的所有断路器。

· 将操作过的所有开关恢复到原来位置。

九、发动机灭火系统的试验和检查

1. 所需设备

三用表。

2. 准备工序

· 拆下后整流罩。

· 拆下两个灭火瓶的引爆管电源插头。

· 从仪表板上拆下灭火系统按钮保护罩。

· 接通直升机电源系统。

3. 检查

进行这些检查需要至少 2 位操作者：

· 一个在座舱里。

· 另一个进行试验。

检查正常的控制操作：

a）依次按下灭火瓶 4 个按钮，当按钮按下时，用三用表检查相应引爆管插头接线柱 A 和 C，B 和 C 或者 1 和 3，2 和 3（采用的插头型号不同其标号方法不同）的电压：

· 左发动机第一个引爆按钮：左灭火瓶黄色引爆管插头。

· 左发动机第二个引爆按钮：右灭火瓶黄色引爆管插头。

· 右发动机第一个引爆按钮：右灭火瓶灰色引爆管插头。

· 右发动机第二个引爆按钮：左灭火瓶灰色引爆管插头。

b）检查（状态、可靠性、锁紧、裂纹、锈蚀）：

· 引爆管电插头。

· 压力开关电插头。

· 灭火瓶（用压力表检查压力，见标牌说明）：

环境温度 /℃	-40	-30	-20	-10	0	+10	+20	+30	+40	+50
压力 /bar±10%	24	26	28.8	32.3	34.6	38	41.8	45.9	50.3	56.2
压力 /MPa±10%	2.4	2.6	2.88	32.3	3.46	3.8	4.18	4.59	5.03	5.62

· 灭火瓶分配管路。

· 后防火墙上的灭火喷嘴。

检查灭火瓶瓶体支架在拐角处是否有裂纹，如果有裂纹，则更换支架。

4. 最后工序

· 断开直升机电源系统（发电机和蓄电池）。

· 在仪表板上安装灭火系统按钮保护罩。

· 拆下试验装置。

· 连接两个灭火瓶引爆管插头。

· 重新装上可卸的后整流罩。

十、旁通活门和油滤堵塞指示器的工作试验

1. 所需设备

模拟滤芯、28V 直流地面电源设备、盛油盘（大约 50L 容量）、放油管。

2. 准备工序

· 将地面电源设备接到直升机插座上。

· 打开主减速器和发动机整流罩。

· 拆下锁线，并放好盛油盘，断开防火墙上的发动机燃油进油管。

· 连接放油管。

3. 试验程序

注：这种试验应依次在每个发动机供油管上进行。

· 拆下滤芯。

· 用模拟滤芯代替原来滤芯，并重新装到油滤中。

· 接通蓄电池开关。

· 接通增压泵开关。

· 依次检查以下各点："油滤堵塞警告灯（FILT）"燃亮；油滤堵塞指示器上的红环全部可以看见。

注：如堵塞警告灯未亮而指示器上的红环已能全部看见，此时用手按压指示帽，堵塞警告灯应点亮。

· 断开增压泵开关。

·断开蓄电池开关。

·拆下模拟油滤组件，并且重新安装原来的滤芯。

·将油滤堵塞指示器复位。

4. 最后工序

·卸下放油管。

·连接和锁紧发动机燃油进口导管，移开盛油盘。

·断开地面电源设备。

·关上发动机整流罩和主减速器整流罩。

十一、消耗油箱低油面开关的工作试验和油量表的校准

1. 所需设备

放油管、带有精度在 0.5% 之内的表的油箱、25L 带刻度的容器、600L 盛油盘。

2. 准备工序

·将直升机调水平。

·打开右侧前检查口盖（以便接近油量表放大器调整螺钉），并打开发动机和主减速器整流罩。

·给直升机供 28V 直流电。

·放上盛油盘，从油箱组中放出部分燃油：在油箱中剩 48kg（即大约 60L）。

·断开防火墙上的发动机供油管，并在同一位置连上放油管。

3. 试验程序（下图）

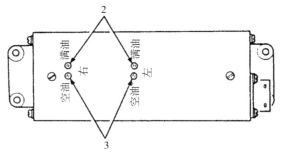

油量表放大器

注：低油面警告灯在增压泵接通前可能亮，但在泵起动后必须熄灭。

· 用直升机的泵连续排出剩余的 60L 燃油，直到消耗油箱低油面警告灯燃亮。

· 将放油管端头放进带刻度的容器，并且连续放油直到泵干运转（空转）。容器中的燃油量必须不少于 20L。

· 拧转油量表放大器上相应的调整螺钉，将油量表指示器调到 0kg 位置上。

· 施加 400kg（大约 506L）的燃油到燃油箱组。

· 拧转油量表放大器上相应的螺钉（2），将油量表指示器调到 400kg 位置上。

注：对于每次调整放大器，允许指示器指针在 2min 内稳定指示。

· 断开放油管。

4. 最后工序

· 放下直升机，使机轮着地。

· 连接并锁紧发动机供油管，移开盛油盘。

· 断开地面电源设备。

· 关上发动机和主减速器整流罩和右侧检查口盖。

十二、电动转输泵和油箱高油面开关的工作试验

1. 所需设备

无。

2. 准备工序

· 加 150kg 燃油到右侧油箱组。

· 加满另一组油箱。

· 接通直升机 28V 直流电源。

3. 试验程序

· 进行一般试验以检查指示灯是否亮。

· 将转输开关扳到右侧。

· 检查："箭头向右（→）"指示灯燃亮；左组油箱油量表指示减少；右组油箱油量表指示增加。

注：当右组油箱加满时，"高油面 (HI.LEV)"指示灯燃亮。

·将转输开关扳到左侧。

·检查："箭头向左（←）"指示灯燃亮；右组油箱油量表指示减少；左组油箱油量表指示增加。

注：当左组油箱加满时，"高油面 (HI.LEV)"警告灯燃亮。

·开关扳到"切断"位置。

4. 最后工序

·断开直升机电源。

十三、液压系统试验

1. 所需设备

无。

2. 系统试验的程序

a）接上电源，警告灯亮：

·"伺服（SERVO）"。

·"液压 1（HYD1）"。

·"液压 2（HYD2）"。

b）按压"伺服 + 行李舱火警（TEST SERVO+CARGO）"试验按钮：

·"伺服（SERVO）"警告灯灭。

3. 旋翼旋转时，警告灯灭：

·"伺服（SERVO）"。

·"液压 1（HYD1）"。

·"液压 2（HYD2）"。

a）试验 7α 信号板按钮。警告灯亮，主警告（ALARM）灯闪亮：

·"伺服（SERVO）"。

·"液压 1（HYD1）"。

·"液压 2（HYD2）"。

b）检查左右系统压力为 6.0 ± 0.2MPa(60 ± 2bar)。

c）按压"伺服 + 行李舱火警（TEST SERVO+CARGO）"试验按钮，警告灯亮：

·"伺服（SERVO）"。

·"行李舱火警（CARGOF）"。

·"过热（O/HEAT）"。

d）试验旁通开关：

·按下 6α 板上的试验按钮 28G，检查 7α 板上的"辅助液压（HYD.AUX）"警告灯亮。

·试验按钮按下，置开关 17G 于"旁通（BY–PASS）"位置，检查"辅助液压（HYD.AUX）"警告灯应熄灭。

·置开关 17G 于"正常（NORM）"位置。

十四、信号灯的试验

1. 所需设备

无。

2. 准备工序

·置"蓄电池接触器 1 和 2（BAT RLY1 和 BAT RLY2）"开关于"断开（OFF）"位置。

·7α 板灯、3Q 板灯及 2G 板灯、17α 灯、3G 灯全应熄灭，当按检查（TEST）和变暗（DIM）按钮时，灯也应全熄灭。

"变暗（DIM）"按钮不通电。

3. 程序

·置"蓄电池接触器 1 和 2（BAT RLY1 和 BAT RLY2）"开关于"接通（ON）"位置。

·大多数灯将燃亮，当按 7α 板上"检查（TEST）"和"变暗（DIM）"按钮时，灯均燃亮，17α 主警告闪亮。

·按下 17α 主警告灯使该灯熄灭。

·压下并保持 7α 板上的"检查（TEST）"按钮，所有 7α 板灯、3Q 和 2G 信号灯将燃亮，并且 17α 主警告灯和 3G 灯将闪亮。

·按下 7α 板上的"变暗（DIM）"按钮，7α 板上警告灯，3Q 板和 2G 板信号灯将变暗，而 17α 主警告灯和 3G 灯亮度将不会改变。

·松开"变暗（DIM）"按钮以获得正常灯光。断开 17α 主警告灯，并释放"检查（TEST）"按钮。

·置"蓄电池接触器 1 和 2（BAT RLY1 和 BAT RLY2）"开关于"断开（OFF）"位置，并且检查是否已恢复到准备工序时的状态。

注：为防止信号板指示灯过热，在地面试验或飞行维护期间信号板应处于"变暗"状态。

十五、主起落架缓冲支柱微动开关的调整

1. 所需设备

三用表、绝缘密封胶、绝缘清漆。

2. 准备工序

·用千斤顶顶起直升机。

·卸下起落架舱门（如果直升机装有舱门）。

3. 微动开关的调整

·检查缓冲支柱是否完全伸出。

·从底座（1）上松开并卸下 3 个螺栓（2），以取下盖子（4）。

·断开微动开关电缆插座。

·将三用表连接到插座上（接线柱 A 和 B）。

·松开锁紧螺母（5）。

·用螺丝刀拧动螺钉（6），以确定微动开关的断开点（A、C）。

·从断开点拧螺钉（6）1～1.5 扣，以保证有适当的间隙。

·把住螺钉（6），拧紧锁紧螺母（5）。

·将 1 滴漆滴到锁紧螺母上。

·在底座（1）上将盖子（4）定好位，并用 3 个在螺纹部分涂绝缘密封胶的螺栓（2）和垫圈（3）固定，用锁线将螺栓和微动开关的螺母锁牢。

·从插座上断开三用表。

重新把微动开关电缆插座连到结构上。

4. 最后工序

·装上起落架舱门。

·调整直升机，使机轮着地。

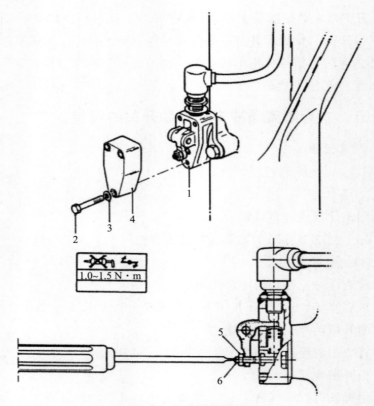

微动开关调整示意图

十六、起落架收起保险联锁系统试验

1. 所需设备

主起落架约束板、三用表、液压千斤顶、28 伏直流地面电源

2. 准备工序

a）给直升机供电。

b）向后倾斜仪表板。

c）在起落架控制开关 1G 的接线柱 1A 和直升机地板之间，连接三用表。

3. 操作工序

注：在试验过程中，不要使液压系统超载运转。检查起落架控制开关是否被安全销固定。

a）在左主起落架上安装主起落架约束板。

b）用千斤顶顶起直升机。

c）检查：

·右主起落架和前起落架缓冲支柱是否伸出。

·前起机轮是否对中并锁住。

d）检查三用表读数是否为零。

e）放下直升机。

f）从左主起落架上拆下主起落架约束板，安装到右主起落架上。

g）用千斤顶顶起直升机。

h）检查：

·左主起落架和前起落架缓冲支柱是否伸出。

·前起机轮是否对中并锁住。

i）检查三用表读数是否为零。

j）放下直升机。

k）从右主起落架上，拆下主起落架约束板。

l）用千斤顶顶起直升机。

m）检查所有缓冲支柱是否伸出。

n）转动前轮，使之与对称中心线成大于2°的偏角。

o）检查三用表的读数是否为零。

p）前轮对中并锁住，将起落架开关置于"应急"位置。

q）检查三用表读数是否为零。

r）前轮对中并锁住和放下直升机，使机轮着地。

s）检查摇臂（2）不与微动开关的接触点（1）接触。

4. 校正工序

a）如果摇臂（2）与微动开关接触点（1）接触，重新调整微动开关。

b）如果在 3d、3i、或 3o 操作工序中，三用表的读数不为零，重新调整有关的微动开关。

c）重新调整微动开关后，重新试验。

d）如果三用表再次偏离零点，更换新的微动开关。

e）在新安装的微动开关上进行试验。

f）如果三用表指针在 3q 操作步骤中偏离零点，更换起落架"应急"开关。

5. 最后工序

a）断开并拆下三用表。

b）放下仪表板并在原来位置锁住。

c）断开地面电源。

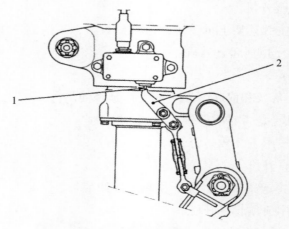

起落架微动开关

十七、备份航姿系统的罗差校正

1. 所需设备

28V 直流地面电源装置。

2. 预备步骤

将 28V 直流地面电源连接到直升机。

3. 步骤

（1）校正要求。

· 直升机在出厂或经大型改装后的第一次试飞前应校正罗差。

· 当直升机更换备份航姿系统处理机时应校正罗差。

· 备份航姿系统在机上通电检查正常后进行罗差校正。

· 校正罗差应在远离钢铁结构的器械、地下电缆、大型金属管道、高大建筑、高压电缆以及其他能够引起地磁变化的杂物 200m 以外的平坦场地进行。

·机上设备（包括随机设备、成品、附件）应配套齐全，并按其正常位置安放。

·在校正前机上所有电子设备应处于正常工作状态。

·在校正没有完成前，不能关电源，否则会产生较大的磁航向误差。

（2）罗差的检查和校正。

准备工作：

·将直升机停放在罗差校正场地上。

·松开直升机的操纵系统，将操纵机构、脚蹬、驾驶杆、总距杆等可动装置置于中立位置（即巡航飞行位置）。

·接通机上所有电子设备，并处于正常工作状态。

罗差校正：

·将备份航姿系统处理机的校正开关拨向"校正"位置（右）。

·接通系统工作电源，等待航向位置指示器开始顺时针转动（永不停止），此时表示系统处于校正状态。

·当系统处于校正状态时，均速推动直升机在罗差校正场地上转动2周（720°）。推动时每周用时应不少于1min。

·将备份航姿系统处理机的校正开关拨向"工作"位置（左）。

·等待航向位置指示器停止转动（航向指向0°方向），此时表示系统已完成校正，断开系统电源。

4. 结束工作

·断开机上全部通电设备的工作电源，拆离地面电源车。

十八、HJG-1A 型组合导航系统磁航向传感器罗差校正

1. 所需设备

无。

2. 校正要求

·直升机在出厂或经大型改装后的第一次试飞前应进行罗差校正。

·在更换磁航向传感器时应进行罗差标定和零偏修正，在更换捷联惯性测量部件时应进行零偏修正。

·罗差校正应在远离钢铁结构的器械、底下地下电缆、大型金属

管道、高大建筑、高压电缆以及其他能够引起地磁变化的杂物 200m 以外的平坦场地进行。

·机上设备（包括随机设备、成品、附件等）应配套齐全，并按其正常位置安放。

·在罗差校正前，机上所有电子设备应能正常工作。

3. 准备工序

·将直升机停放在罗盘校正场地上。

·接通机上所有电子设备，并处于工作状态。

4. 罗差校正

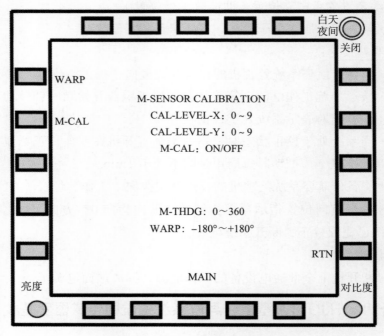

组合导航磁传感器罗差校准页面

M–THDG：磁传感器磁航向角　WARP：磁传感器零偏修正值

CAL–LEVEL–X/Y：X 轴 /Y 轴修正级别

M–CAL：标定启动键（2min 后自动结束，标定过程中再次按下该键可中断并停止标定）。

（1）罗差标定。

分别按多功能显示器（MFCD）上 D3、R1、L1，进入导航状态的控制页面，再按 M–CAL 进入磁航向传感器罗差校正页面，如上图所示。按 M–CAL 键启动标定，页面显示 M–CAL：ON，同时匀速推动直升机在 2min 内转动一周，组合导航系统 2min 完成标定，标定结束后页面显示 M–CAL：OFF，同时显示标定结果，标定结构分为 0~9 级，由 CAL LEVEL–X/Y：0~9 表示，9 为最好，小于 8 时须重新标定。

（2）零偏修正。

罗差标定完成后，将直升机对称中心线对准罗盘场地上的 0° 方位（磁航向正北）。在多功能显示器磁航向传感器罗差校正页面状态下，读取磁航向传感器输出的磁航向角度值，并计算出其零偏值，按 WARP 键，在多功能键盘上输入磁航向传感器当前的零偏修正值。例如：如果磁航向传感器磁航向值（M–HDG）为 2°，则输入 2；如果磁航向传感器磁航向值（M–HDG）为 358°，则输入 –2°，再按"ENT"键确认，即完成磁航向传感器零偏修正。

5. 最后工序

· 将 MFK 上 INS 和 IDCS 工作状态开关置"OFF"位，将 ICP 上的"HNS"开关置于"OFF"位置。

· 关闭机上全部通电设备的电源。

· 撤离地面电源。

十九、发动机超应急功率控制系统试验

1. 所需设备

28V 直流地面电源装置。

2. 准备工序

· 记录安装在发动机功率状态检测记录盒上的工作时数，插上地面电源插头。

· 置"蓄电池接触器开关（49P、50P）"于接通位，则 7α 板上"蓄电池接触器 1（BAT.RLY1）"、"蓄电池接触器 2（BAT.RLY2）""发电机 1（GEN1）""发电机 2（GEN2）""汇流条相连 (BUS–CPL)"指示灯燃亮。

3. 试验

接通音响警告开关 37E，将机通控制盒上第 7 通道旋钮调至音量最大位置。

·按正常程序起动一台发动机，Ng 转速达到 40％时松开起动按钮，推油门进行加速直至飞行位置，并使扭矩略大于 25％，检查仪表板上两发超应急功率警告灯燃亮，耳机内能听到连续的警告"锣"声，然后油门拉回至地面慢车位置，使扭矩＜ 25％，检查两发超应急功率警告灯应熄灭，耳机内警告"锣"声停止。

·起动另一台发动机，前推油门杆加速直至飞行位置，并使此台发动机扭矩比另一发动机扭矩大 25％以上，检查仪表板上两发超应急功率警告灯应燃亮、耳机内能听到连续的警告"锣"声，然后拉回油门杆，令此发动机停车。

·断开 12 α 板上发动机超应急控制开关 121E，前推仍在运转的发动机油门杆至飞行位置，并使扭矩＞ 25％，检查两发超应急功率警告灯不亮，耳机内听不到连续的警告"锣"声。

·后拉油门杆使发动机停车。打开主减整流罩，检查发动机功率状态检测记录盒上记录的发动机工作时间应与实际工作时间 (Ng ＞ 50％始算) 相符，警告旗没有翻转。

4. 恢复

·拔下地面电源插头；

·关闭主减整流罩；

·接通 12α 板上的发动机超应急控制开关 121E，并用锁丝重新铅封。

注：将直升机系留在车台上进行试验。

Overall Requirements and Standards for Helicopter Maintenance

1. Condition (Exterior)

Conduct visual check to see if there is any deformation, damage, fracture, abrasion, scratch, erosion, dust, dirt, overheated place and so on.

2. Fixation

Conduct visual check on mounting points (including conditions of locking devices and markings of self-locking nuts). Make sure each mounting point is tightly fastened, well secured and well protected by shock absorbers. Hand check can also be used during visual check (use hands to check parts and components that might be malfunctioning). Maintainers could use hands to gently shake target parts or components and check if there is any abnormal gap or clearance in the movement.

3. Cables and cable plugs

(1) Cables and cable plugs shall be clean, well insulated and well wrapped against abrasion, oil/fuel and water. Anti-wave sleeves shall be intact and well grounded.

(2) Harnesses shall be well aligned, fastened and insulated. They shall be wrapped to avoid abrasion with other parts or bulkheads of the helicopter. The bending angles of wires shall exceed 90°. Proper length shall be reserved for movable wires.

(3) Pins and bolts shall be well installed. Nuts shall be well fastened and secured according to the standards.

(4) Electric contacts such as pins, jacks and negative wires shall be clean and well connected. Fracture of a bonding jumper (or a harness of bonding jumpers) shall not exceed 1/3 of the total diameter.

4. Pipes

Conduct visual check to see if there is any leakage in the connecting points of pipes and sleeves. Make sure the joints are well fastened by clamps. Pipes shall not be violently bended or abraded. Note: please pay special attention to the outlets and joints of hydraulic, fuel and oil systems and the pipes in the pitot-static system during such checks.

Direction for Use

(1) The manual consists of three parts: exterior check, power-on check and special check. In exterior check, the helicopter is divided into 11 stations. The contents of power-on check are summarized based on the requirements for power-on check on each instrument and device and the practice of units.

(2) The check on each station is illustrated with a picture or pictures in which the names of relative devices are marked. The check on each device shall proceed according to the number marked in front each device.

(3) All devices marked in each picture shall be checked in the current station, unless they are marked by white letters or frames.

(4) Please refer to the explanations after each picture to see the basic working principles, parameters and requirements for the check on each device. The contents above will be explained one by one based on the number marked in front of each device. To facilitate trouble-shooting and wire/line-hunting, each device is noted with the type and electric symbol (there are no independent electric symbols for devices on engines as they are treated as a whole).

(5) Due to limited space, only the key points to check will be listed for each device. Please refer to overall requirements and standards for helicopter maintenance to see specific requirements.

Chapter 1　Exterior Check

Check Stations of Z-9 Helicopter

Station 1: Exterior of cockpit

Station 2: Left side of fuselage

Station 3: Left side of MGB (main gear box)

Station 4: Left side of engine

Station 5: Left side of tail boom and tail rotor hub

Station 6: Right side of tail boom and tail gear box

Station 7: Right side of fuselage and bottom structure

Station 8: Right side of MGB

Station 9: Right side of engine

Station 10: Cockpit

Station 11: Interior of cargo bay

Station 1

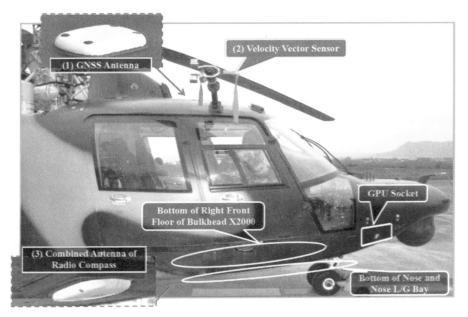

Station 1: Exterior of Cockpit (Right Side of Nose)

(1) GNSS Antenna (43S): It receives satellite positioning signals and sends the signals to GNSS receiver inside the inertial measuring unit.

Key Points to Check: appearance and installation.

(2) Velocity Vector Sensor GSS-1A (102F): It senses total pressure, static pressure, total temperature of the induced airflow and the relative azimuth angle to helicopter coordinates. The data above will further be sent to ADC (air data computer).

Key Points to Check: appearance, installation and flexibility.

(3) Combined Antenna of Radio Compass ZT-9 (81R): It receives signals from beacons, radio stations or navigation beacons and sends the signals to the receiver of radio compass WL-9-1.

Key Points to Check: appearance and installation.

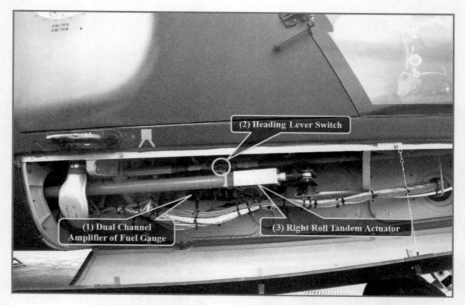

Bottom of Right Front Floor of Bulkhead X2000

(1) Dual Channel Amplifier of Fuel Gauge FUC-46 (2Q): It's a component of BUC-46 capacitive fuel gauge. It processes the change in the capacitance of the fuel gauge sensor and outputs DC voltage in direct proportion to the fuel quantity to the indicator.

Key Points to Check: appearance, installation, cable plug(s) and cable(s).

(2) Heading Lever Switch CHG-1 (28C): As a rigid linkage, it transmits the force applied on the pedals to tail rotor through an actuator so that the pilot could change the heading. As an electric switch, when the CSAS (control stability augmentation system) is engaged, it could activate the corresponding electric conversion (cancel anchoring) in the CSAS and affect flight control status based on whether the corresponding switch of the heading channel is connected or not and how much force it senses.

Key Points to Check: appearance, installation, cable plug(s) and cable(s).

(3) Right Roll Tandem Actuator DCD-10 (24C): It receives signals from the amplifier, executes control on retraction and extension and feeds back signals to indicate the position of the actuator.

Key Points to Check: appearance, installation, cable plug(s) and cable(s).

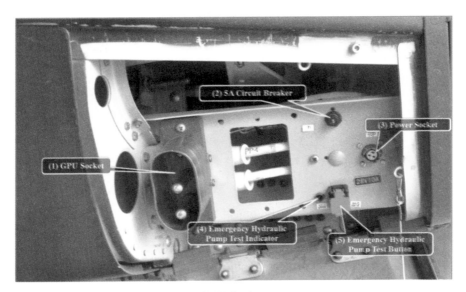

GPU Socket

(1) GPU Socket (23P): It is connected to a ground DC power source and supplies power to heliborne DC power system.

Key Points to Check: appearance, installation and cable(s).

(2) 5A Circuit Breaker (11P): It is connected to the ground power over-voltage protection plug-in unit. When the over-voltage protection circuit is short-circuited, the breaker will be disconnected and ground power will be cut off from heliborne power network.

Key Points to Check: appearance and installation.

(3) Power Socket (90P): It is a standby DC power socket.

Key Points to Check: appearance and installation.

(4) Emergency Hydraulic Pump Test Indicator (24G): If the emergency hydraulic pump works normally in a ground test, the indicator will light up.

Key Points to Check: appearance and installation.

(5) Emergency Hydraulic Pump Test Button (20G): It is used to test the emergency hydraulic pump in a ground test.

Key Points to Check: appearance, installation and flexibility.

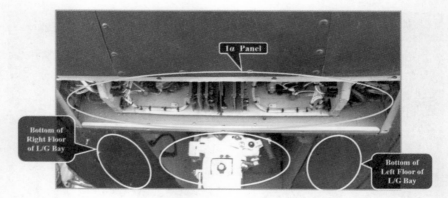

Bottom of Nose

1α Panel: As the main power distribution box, it is equipped with 4 contactors, 2 short-circuit protection plug-in units, 2 reverse current protectors, 2 diode plug-in units, 2 reverse current breakers, 2 shedding contactors (one for left system and the other for right system), 2 ground power relays, 2 short-circuit detection relays and 1 ground power over-voltage protection plug-in unit.

Key Points to Check: appearance and installation.

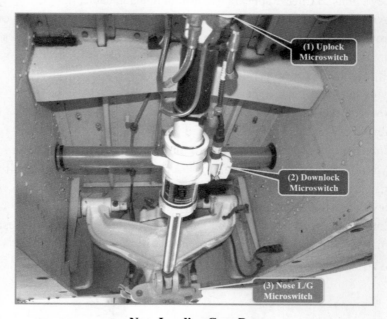

Nose Landing Gear Bay

(1) Uplock Microswitch (6G): It is locked and disconnects the retraction circuit when the landing gear is fully retracted. It is unlocked when the landing gear is being retracted, extended and fully extended, setting the retraction circuit ready.

Key Points to Check: appearance, installation, cable plug(s) and cable(s).

(2) Downlock Microswitch (7G): It is locked and disconnects the extension circuit when the landing gear is fully extended. It is unlocked when the landing gear is being extended, retracted and fully retracted, setting the extension circuit ready.

Key Points to Check: appearance, installation, cable plug(s) and cable(s).

(3) Nose L/G Microswitch (5G): As the microswitch of the shock-absorbing leg, it is engaged when the shock-absorbing leg is compressed and set in neutral position at the same time (when the helicopter is parked on the ground) to prevent the landing gear from being retracted.

Key Points to Check: appearance, installation, cable plug(s) and cable(s).

Bottom of Left Floor of Nose Landing Gear Bay

(1) Fire Control Box HKH-8A (7E): When the area temperature of left engine exceeds the warning value, the relay is connected and sends warning

signals. When the fire detection circuit is disconnected or grounded, the control box sends failure signals.

Key Points to Check: appearance, installation, cable plug(s) and cable(s).

(2) Connection Box (30α): It consists of the wiring of systems including the hydraulic system, communication system, anti-icing and rain-draining system, landing gear, illumination system, power distribution system and engine control system.

Key Points to Check: appearance, installation, cable plug(s) and cable(s).

(3) Relay Box (23α): It consists of the relays of systems includin the fire extinguisher system, cabin door warning system, hydraulic system and extension/retraction system of the landing gear.

Key Points to Check: appearance, installation, cable plug(s) and cable(s).

(4) Plug (48A): They are the control potentiometers of left and right collective pitch levers and the adapter plug for emergency jettison circuit.

Key Points to Check: appearance, installation, cable plug(s) and cable(s).

(5) Connection Box (31α): It is the connection box of the illumination system and power distribution devices.

Key Points to Check: appearance, installation, cable plug(s) and cable(s).

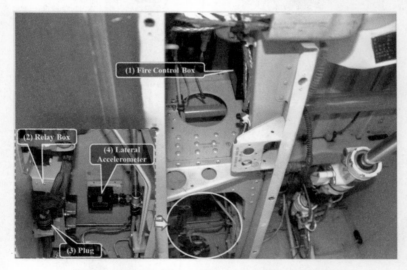

Bottom of Right Floor of Nose Landing Gear Bay

(1) Fire Control Box HKH-8A (7E): When the area temperature of right engine exceeds the warning value, the relay is connected and sends warning signals. When the fire detection circuit is disconnected or grounded, the

control box sends failure signals.

Key Points to Check: appearance, installation, cable plug(s) and cable(s).

(2) Relay Box (24α): It consists of the control relays of systems including the fire extinguisher system, hydraulic system, anti-icing and rain-draining system and illumination system.

Key Points to Check: appearance, installation, cable plug(s) and cable(s).

(3) Plug (14α): It is the cable plug of the cyclic pitch lever of the pilot.

Key Points to Check: appearance, installation, cable plug(s) and cable(s).

(4) Lateral Accelerometer GJ-10B (58F): It is used to sense lateral acceleration of the helicopter. It will convert the change in the acceleration into 400Hz signals and input them to the heading channel of the CSAS to realize the coordinated turn of the helicopter.

Key Points to Check: appearance, installation, cable plug(s) and cable(s).

Front of Nose

(1) Windshield Wiper: It is used to wipe off the rainwater and snowflakes on the windshield. Its high speed is not less than 60 times/min, with low speed not less than 40 times/min. The maximum mechanical angle is 45°. Its DC power is 27V and the motor power is 141W. It is not allowed to use windshield wipers on dry windshields.

Key Points to Check: appearance, installation, wiper blades and pressure.

(2) Front Antenna of Transponder (72S): It receives the identification/
interrogation signals (in general mode, mode 1, mode 2, mode 3 or mode
4) and non-identification/interrogation signals (batch number and altitude)
from ground, shipboard or other airborne interrogators (or interrogator
transponders) and then sends such signals to the transponder.
Key Points to Check: appearance and installation.

(3) Pitot Tube: It senses and transmits the value of total pressure to the air
speed indicator and ADC.
Key Points to Check: appearance, installation and pipe.

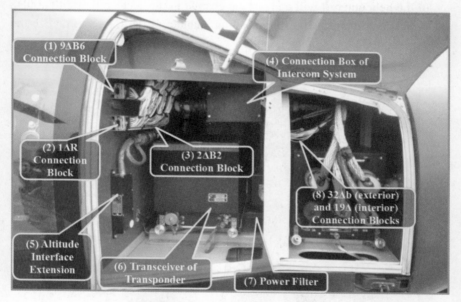

Right Side of Radio Equipment Bay

(1)(2)(3)(8) Connection Block: 9ΔB6: connection block of the navigation
system and the CSAS; 1ΔR: line bank of the intercom system; 2BΔ2:
connection block of the power system and engine control system; 32Δb:
connection box of the CSAS, fuel system, hydraulic system and indication
system; 19Δ: connection box of the anti-icing and rain-draining systems.
Key Points to Check: appearance, installation and cable(s).

(4) Connection Box of Intercom System (22R): It is composed of receiving,
transmitting, amplifying, power filtering and warning circuits. Sockets J1,
J2 and J3 are connected to the control box. J6 is connected to heliborne
radio equipment and the power system. J4 and J5 are not in use.

Key Points to Check: appearance, installation, cable plug(s) and cable(s).

(5) Altitude Interface Extension 9A1 (79S): It receives non-calibrated barometric altitude signals from the altimeter and the HADS (Helicopter Air Data System). There are ARINC-429 interface of the HADS and RS-422 altitude interface.

Key Points to Check: appearance, installation, cable plug(s) and cable(s).

(6) Transceiver of Transponder K/LKB015A (70S): It receives and decodes the identification/ interrogation signals (in general mode, mode 1, mode 2, mode 3 or mode 4) and non-identification/interrogation signals (batch number and altitude) from ground, shipboard or other airborne interrogators (or interrogator transponders). Then it sends responsive identification/non-identification signals (batch number and altitude) or special responsive signals (in special modes 1 to 5) so that the interrogator could identify the target and record its batch number and altitude.

Key Points to Check: appearance, installation, cable plug(s) and cable(s).

(7) Power Filter 25D (75S): It filters power signals.

Key Points to Check: appearance, installation, cable plug(s) and cable(s).

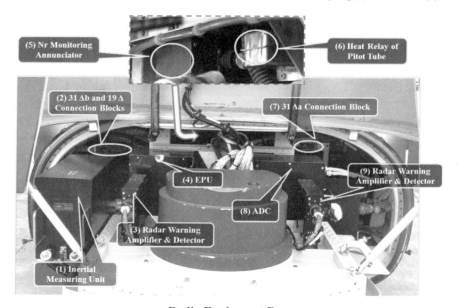

Radio Equipment Bay

(1) Inertial Measuring Unit HJG-1A (41S): It receives and processes acceleration signals and gyroscope signals, and transmits the processed

signals to the IDCS (Integrated Display and Control System), IFCS (Integrated Fire Control System) and navigation system.

Key Points to Check: appearance, installation, cable plug(s) and cable(s).

(2) 31Δb and 19Δ Connection Blocks: They are the connection blocks of the anti-icing and rain-draining system.

Key Points to Check: appearance, installation and cable(s).

(3)(9) Radar Warning Amplifiers & Detectors ARW9503-1 (5T and 8T): They receive radar signals from radar warning antennas and transmit amplified and detected signals to the signal analyzer.

Key Points to Check: appearance, installation, cable plug(s) and cable(s).

(4) EPU EFC-2-02 (204E): Engine parameter acquisition unit, or EPU in short, provides DC power at ± 15V to the sensors (of the MGB, hydraulic system and engines) in need, acquires necessary data (about the MGB, hydraulic system as well as pressure, temperature, rotation speed, torque and status of engines), and receives parameters including wheel load, power, atmospheric temperature and flight altitude from related sensors and devices. Then it transmits the calculated and processed data to related devices (the EPD and warning indicators).

Key Points to Check: appearance, installation, cable plug(s) and cable(s).

(5) Nr Monitoring Annunciator XJZ-6 (11E): Working together with 832E, it monitors Nr and generates high-pitch and low-pitch warning sounds.

Key Points to Check: appearance, installation, cable plug(s), cable(s).

(6) Heat Relay of Pitot Tube (4F): It connects and disconnects the heat circuit of the pitot tube.

Key Points to Check: appearance, installation and cable(s).

(7) 31Δa Connection Block: It is a connection block.

Key Points to Check: appearance, installation and cable(s).

(8) ADC XSC-5A (101F): Equipped for the integrated task system, it receives such data as total pressure, static pressure and total temperature of the induced airflow as well as the relative azimuth to helicopter coordinates. Also, it receives data including the vertical overload, rotational angular velocity and helicopter weight from the bus. After processing and calculating these data, it produces parameters needed by the helicopter, such as 3-axis air speed, pressure altitude, vertical speed, total temperature, rotor induced airspeed, sideslip angle and angle of attack. The system is capable of built-in-test (BIT).

Key Points to Check: appearance, installation, cable plug(s) and cable(s).

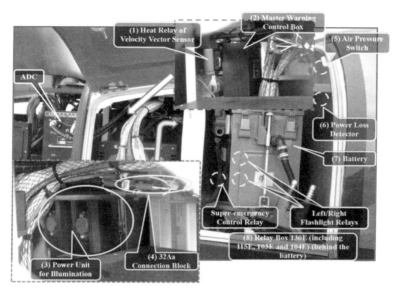

Left Side of Radio Equipment Bay

(1) Heat Relay of Velocity Vector Sensor JKA-52B (106F): It connects and disconnects the heat circuit of the velocity vector sensor.
Key Points to Check: appearance, installation and cable.

(2) Master Warning Control Box KZH-55A (11α): It engages the corresponding warning indicators based on the signals from sensors and provides power to flash the main alarm.
Key Points to Check: appearance, installation, cable plug(s) and cable(s).

(3) Power Unit for Illumination (50L): It provides power to illuminating devices in the cockpit.
Key Points to Check: appearance, installation, cable plug(s) and cable(s).

(4) 32Δa Connection Block: It is an adapter module for the CSAS, fire extinguisher system, fuel system, hydraulic system, indication system and landing gear system.
Key Points to Check: appearance, installation and cable(s).

(5) Air Pressure Switch (4G): It is connected to the landing gear warning indicator on the instrument panel. If the landing gear is not fully extended and locked when air speed is lower than 101.92km/h (55kn), it will send warning signals.
Key Points to Check: appearance, installation, cable plug(s) and cable(s).

(6) Power Loss Detector XJZ-2A (32E): It receives signals from the Nf sensor.

When Nf is up to the warning limit, the detector is grounded and the corresponding power loss indicator lights up (Warning limit: 313r/min. Precision: ± 1%). Warning signals will be transmitted (the indicator is on) when Nf is up to the warning limit until the engine shuts down.

Key Points to Check: appearance, installation and cable(s).

(7) Battery (22P): It is the emergency and backup DC power source at a rated capacity of 36Ah and rated voltage of 24V.

Key Points to Check: appearance, installation, cable plug(s) and cable(s).

(8) Relay Box (130E, including 115E, 103E and 104E): If one engine fails and the torque difference between two engines ΔC exceeds 25%, it connects the circuit to light up the super-emergency power indicator (on top of the EPD).

Key Points to Check: appearance, installation, cable plug(s) and cable(s).

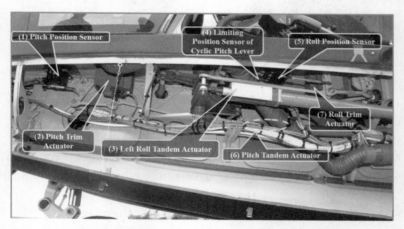

Bottom of Left Front Floor of Bulkhead X2000

(1) Pitch Position Sensor GE-23 (71C): It senses and transmits the signals that indicate longitudinal angle position to the computer.

Key Points to Check: appearance, installation, cable plug(s) and cable(s).

(2) Pitch Trim Actuator DCD-9 (21C): It receives computer instructions and establishes an anchoring point for the control linkage. There is a damper. The actuator will be anchored when the electromagnetic clutch is not engaged (not switched on). A load sensitive switch will release the lever force when certain force is applied on the cyclic pitch lever.

Key Points to Check: appearance, installation, cable plug(s) and cable(s).

(3) Left Roll Tandem Actuator DCD-10B (25C): It receives signals from the amplifier, executes control on retraction and extension and feeds back

signals to indicate the position of the actuator.

Key Points to Check: appearance, installation, cable plug(s) and cable(s).

(4) Limiting Position Sensor of Cyclic Pitch lever: It sends signals to the corresponding indicator "Limit" when the cyclic pitch lever is on left or right limiting position.

Key Points to Check: appearance, installation, cable plug(s) and cable(s).

(5) Roll Position Sensor GE-23 (73C): It senses and transmits the signals that indicate horizontal angle position to the computer.

Key Points to Check: appearance, installation, cable plug(s) and cable(s).

(6) Pitch Tandem Actuator DCD-10 (23C): It receives signals from the amplifier, executes control on retraction and extension and feeds back signals to indicate the position of the actuator.

Key Points to Check: appearance, installation and cable(s).

(7) Roll Trim Actuator DCD-9 (27C): It receives computer instructions and establishes an anchoring point for the control linkage. There is a damper. The actuator will be anchored when the electromagnetic clutch is not engaged (not switched on). A load sensitive switch will release the lever force when certain force is applied on the cyclic pitch lever.

Key Points to Check: appearance, installation, cable plug(s) and cable(s).

Station 2

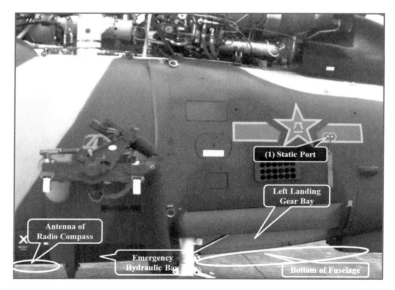

Station 2: Left Side of Fuselage

(1) Static Port: It senses and transmits static pressure to the total-static system and air data calculator.

Key Points to Check: appearance and blockage.

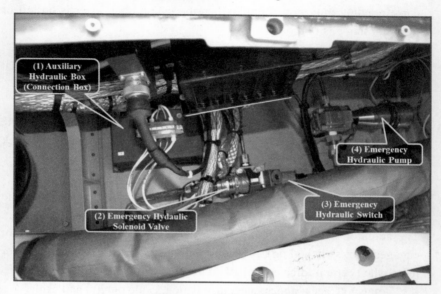

Emergency Hydraulic Bay

(1) Auxiliary Hydraulic Box (Connection Box) (19G): It is the connection box of the auxiliary hydraulic system.

Key Points to Check: appearance, installation, cable plug(s) and cable(s).

(2) Emergency Hydraulic Solenoid Valve YDK-12 (26G): When powered on, it opens the pipes of the emergency hydraulic system.

Key Points to Check: appearance, installation, cable plug(s), cable(s) and leakage.

(3) Emergency Hydraulic Switch YCG-3 (23G): It lights up an indicator when the emergency hydraulic pressure exceeds 95 bars to indicate that the emergency system is working.

Key Points to Check: appearance, installation, cable plug(s), cable(s) and leakage.

(4) Emergency Hydraulic Pump YCB-0.25 (21G): It supplies hydraulic oil to main and auxiliary hydraulic systems when the main hydraulic system fails.

Key Points to Check: appearance, installation, cable plug(s), cable(s) and leakage.

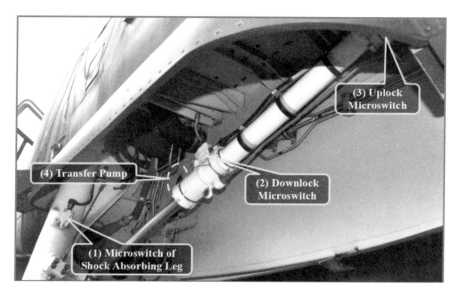

Left Landing Gear Bay

(1) Microswitch of Shock-absorbing Leg (9G): It senses if the landing gear has left the ground so as to prevent it from being retracted when the shock-absorbing leg is not fully extended (or not off the ground).

Key Points to Check: appearance, installation, cable plug(s) and cable(s).

(2) Downlock Microswitch (13G): It disconnects the circuit of the hydraulic solenoid valve and sets the retraction circuit ready when the landing gear is fully extended.

Key Points to Check: appearance, installation, cable plug(s) and cable(s).

(3) Uplock Microswitch (11G): It disconnects the circuit of the hydraulic solenoid valve and sets the extension circuit ready when the landing gear is fully retracted.

Key Points to Check: appearance, installation, cable plug(s) and cable(s).

(4) Transfer Pump RXB-4 (20Q): It transfers fuel from one tank group to the other.

Key Points to Check: appearance, installation, cable plug(s), cable(s) and leakage.

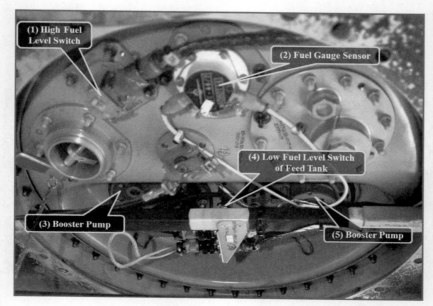

Bottom of Fuselage (Left)
(Right side is the same but without the high fuel level switch)

(1) High Fuel Level Switch XU-24 (9Q): During fuel transfer, it will be connected and light up indicator "HI.LEV" when the fuel tank receiving fuel is full.

Key Points to Check: appearance, installation, cable plug(s), cable(s) and leakage.

(2) Fuel Gauge Sensor GUC-46/3 (5Q): It is a component of the capacitive fuel quantity measuring unit, providing data of the quantity of the fuel stored in the fuel tank.

Key Points to Check: appearance, installation, cable plug(s), cable(s) and leakage.

(3)(5) Booster Pumps RLB-12 (7Q and 69Q): They raise pressure in fuel pipes.

Key Points to Check: appearance, installation, cable plug(s), cable(s) and leakage.

(4) Low Fuel Level Switch of Feed Tank XU-24/3 (11Q): It is connected and lights up indicator "FUEL.Q" on 3Q panel when there is less than 18L of fuel in the feed tank. Indicator "FUEL.Q" on 7α panel also lights up. It also connects the logic circuit 11α of the warning system and lights up

indicator "ALARM" on the instrument panel.

Key Points to Check: appearance, installation, cable plug(s), cable(s) and leakage.

Station 3

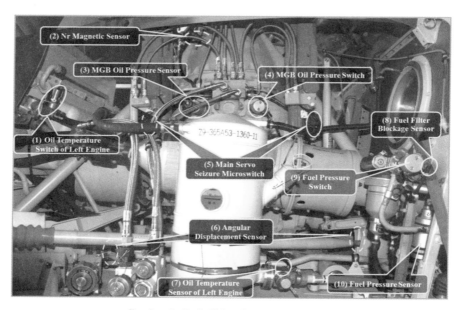

Station 3: Left Side of Main Gear Box

(1) Oil Temperature Switch of Left Engine TW-9 (47E): It senses the oil temperature of the left engine. The warning indicator lights up when the temperature exceeds 130℃ .

Key Points to Check: appearance, installation, cable plug(s), cable(s) and leakage.

(2) Nr Magnetic Sensor (832E): It senses rotation rate of the rotor and transmits warning signals to the horn warning annunciator and 7α panel. The low-pitch warning sound is 85Hz-172.5Hz when Nr is between 170r/min and 345r/min; the high-pitch warning sound is 190Hz when Nr is over 380r/min.

Key Points to Check: appearance, installation, cable plug(s) and cable(s).

(3) MGB Oil Pressure Sensor CY-YZ-0101-0.3 (17E): It measures the oil pressure of the MGB. It is connected to the EPU.

Key Points to Check: appearance, installation, cable plug(s), cable(s) and leakage.

(4) MGB Oil Pressure Switch (16E): It lights up the warning indicator when the oil pressure drops below 0.08MPa.

Key Points to Check: appearance, installation, cable plug(s) and cable(s).

(5) Main Servo Seizure Microswitch: It is a warning device to indicate the failure of the rotary valve. It sends warning signals and lights up indicator "SERVO" to warn the pilot.

Key Points to Check: appearance, installation, cable plug(s) and cable(s).

(6) Angular Displacement Sensor: It senses and transmits the position signals of the transmission system to the flight data system. There are no electrical symbols for additional retrofitted devices.

Key Points to Check: appearance, installation and cable(s).

(7) Oil Temperature Sensor of Left Engine GWR-2A (21E): It senses and converts the oil temperature of the left engine into electric signals, and transmits the signals to the EPU for processing and display.

Key Points to Check: appearance, installation, cable plug(s), cable(s) and leakage.

(8) Fuel Filter Blockage Sensor (18Q): Indicator "FILT" on 3Q panel and indicator "FUEL" on 7α panel light up when the pressure difference is over 140 ± 20mbar.

Key Points to Check: appearance, installation, cable plug(s), cable(s), and leakage.

(9) Fuel Pressure Switch (14Q): It senses the fuel pressure at the outlet of the low-pressure pump. When the pressure is lower than 0.2bar, it lights up indicator "FUEL" on 7α panel and indicator "FUEL.P" on 3Q panel.

Key Points to Check: appearance, installation, cable plug(s), cable(s) and leakage.

(10) Fuel Pressure Sensor CY-YZ-0101-0.2 (17Q): Its resistance value changes with the fuel pressure. The sensor transmits the corresponding voltage signals to the EPU and the EPD will display the processed signals. Range of measurement: 0-0.2MPa.

Key Points to Check: appearance, installation, cable plug(s), cable(s) and leakage.

Station 4

Station 4: Left Side of Engine (Left Engine Bay)

(1) Super-emergency Solenoid Valve: If one engine fails and the torque difference ∆C between two engines exceeds 25% in flight, it will be connected to supply super-emergency fuel. The corresponding super-emergency power warning indicator on top of the EPD will light up at the same time.
Key Points to Check: appearance, installation, cable plug(s), cable(s) and leakage.

(2) Air Bleed Valve: It is opened during engine start process to adjust air inflow and prevent engine surges.
Key Points to Check: appearance, installation, cable plug(s) and cable(s).

(3) Start Solenoid Valve RDK-16: It is a fuel injector solenoid valve that provides fuel to the combustion chamber at the initial phase of the start-up process. When the fuel pressure reaches 0.24MPa (2.4bar), it begins to inject air instead of fuel.
Key Points to Check: appearance, installation, cable plug(s) and cable(s).

(4) Engine Oil Temperature Switch HJK-3: It lights up warning indicator "ENG.1" when the oil pressure drops below 1.3bar.
Key Points to Check: appearance, installation, cable plug(s), cable(s) and leakage.

(5) Engine Oil Pressure Sensor CY-YZ-0101-0.8L: The piezoresistive sensor contains a bellow connected to induction coils. It measures the oil pressure of each engine. The measurement range of pressure is 0-0.8MPa. The data

will be transmitted to the EPU for processing and display.

Key Points to Check: appearance, installation, cable plug(s), cable(s) and leakage.

(6) Ignition Plug: It ignites the mixture of fuel and air.

Key Points to Check: appearance, installation and cable(s).

(7) Metal Chip Detector: It monitors the degree of abrasion in the bearing cavities of the engines. When excessive metal chips attracted fill the loop gap, the circuit will be connected and the indicator in the cockpit will light up to warn the flight crew.

Key Points to Check: appearance, installation, cable plug(s), cable(s) and leakage.

(8) Nf Sensor GZ-20: It measures the rotation rate of the free turbine and transmits the data to the EPU for engine power loss warning.

Key Points to Check: appearance, installation, cable plug(s), cable(s) and leakage.

(9) Overspeed Cut-off Solenoid Valve: When Nf is over 123.1%+2.5% of the rated value, the two sensors will detect the overspeed and the valve will shut down the engine.

Key Points to Check: appearance, installation, cable plug(s), cable(s) and leakage.

(10) Overspeed Sensor: It senses Nf, sends overspeed warning signals and shuts down the engine.

Key Points to Check: appearance, installation, cable plug(s), cable(s) and leakage.

Station 5

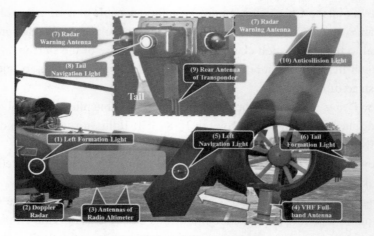

Station 5: Left Side of Tail Boom

(1)(6) Left Formation Light BD-12 (147L) and Tail Formation Light BD-12 (144L): They provide information needed by pilots of other helicopters in a formation flight. There are an IFR mode and a normal mode.
Key Points to Check: appearance and installation.

(2) Doppler Radar 777 (22S): It is a type of radar that uses Doppler effect to identify the location and velocity of the target.
Key Points to Check: appearance and installation.

(3) Antennas of Radio Altimeter: The transmitting antenna（rear, 14S）transmits FM CW signals and the receiving antenna (front, 13S) receives reflected Signals.
Key Points to Check: appearance and installation.

(4) VHF Full-band Antenna TKR 128 (43R): It is a VHF full-band antenna with the working frequency bands of 30MHz-87.975MHz and 108MHz-155.975MHz. It is mainly used for plain/secure communications between helicopters and between helicopters and the ground.
Key Points to Check: appearance and installation.

(5)(8) Left Navigation Light HD-13 (15L) and Tail Navigation Light WD-3A (16L): The left navigation light is in red (4.2W), with the right one in green (4.2W) and the tail one in white (2.4W). They are used to locate the helicopter and indicate the heading. There are an IFR mode and a normal mode.
Key Points to Check: appearance and installation.

(7) Radar Warning Antennas (left 3S and right 4S): They intercept and transmit detective signals of radars to receivers on both sides of the helicopter.
Key Points to Check: appearance and installation.

(9) Rear Antenna of Transponder (73S): It receives and decodes the identification/interrogation signals (in general mode, mode 1, mode 2, mode 3 or mode 4) and non-identification/interrogation signals (batch number and altitude) from ground, shipboard or other airborne interrogators (or interrogator transponders). Then it will send responsive identification/non-identification signals (batch number and altitude) or special responsive signals (in special modes 1 to 5) so that the interrogator could identify the target and record its batch number and altitude.
Key Points to Check: appearance and installation.

(10) Anticollision Light FZD-20 (18L): There are a normal mode and an IFR mode. In normal mode, the helicopter could be located in a long-distance flight and ground operation. In IFR mode, the helicopter could be identified by other helicopters in flight or in ground operation during a

long-distance flight so as to prevent collision. Its operating voltage is 28V and flash rate is 60 times/min

Key Points to Check: appearance and installation.

Station 6

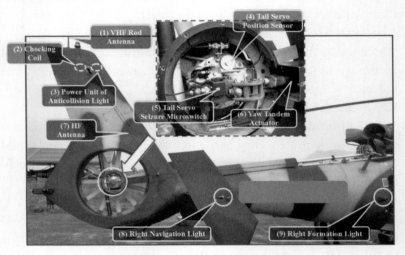

Station 6: Right Side of Tail Boom

(1) VHF Rod Antenna (52R): It receives RF (radio frequency) signals and transmits carrier signal. The frequency band is 30–87.975MHz.

Key Points to Check: appearance and installation.

(2) Chocking Coil (97L) (3) Anticollision Light Power Unit KZH-125 (13L): They provide the anticollision light with power and trigger flashes in bright red to avoid collision with other airplanes or aircrafts.

Key Points to Check: appearance, installation, cable plug(s) and cable(s).

(4) Tail Servo Position Sensor (30FA): It transmits signals to the flight data system to indicate the position of the tail servo.

Key Points to Check: appearance, installation, cable plug(s) and cable(s).

(5) Tail Servo Seizure Microswitch (11D): It is a warning device that indicates the failure of the rotary valve. It lights up indicator "SERVO" to warn the flight crew.

Key Points to Check: appearance, installation, cable plug(s) and cable(s).

(6) Yaw Tandem Actuator DCD-10A (26C): It receives signals from the amplifier, executes control on retraction and extension and feeds back signals to indicate the position of the actuator.

Key Points to Check: appearance, installation, cable plug(s) and cable(s).

(7) HF (Wire Rope) Antenna (13R): It receives short-wave RF signals and transmits carrier waves.

Key Points to Check: appearance and installation.

(8) Right Navigation Light HD-13 (17L) and (9) Right Formation Light BD-12 (146L): Their functions and parameters are the same as those of the devices on the left side.

Key Points to Check: appearance and installation.

Station 7

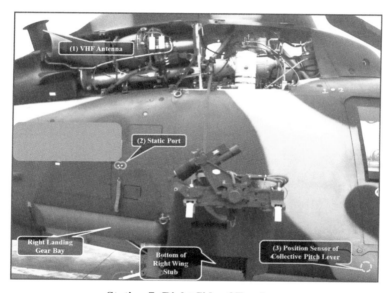

Station 7: Right Side of Fuselage

(1) VHF Antenna (55R): It receives RF signals and transmits carrier signals. Its working frequency bands are 108−155.975MHz, 156−173.975MHz and 225−399.975MHz.

Key Points to Check: appearance and installation.

(2) Static Port: They measure and transmit static pressure to the barometric altimeter and air data calculator.

Key Points to Check: appearance and blockage.

(3) Position Sensor of Collective Pitch lever GE-14 (66C): It senses the position of the collective pitch lever and transmits position signals to the computer.

Key Points to Check: appearance, installation, cable plug(s) and cable(s).

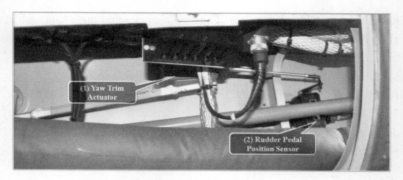

Bottom of Right Wing Stub

(1) Yaw Trim Actuator DCD-9A (22C): As a parallel actuator, it receives instructions of the amplifier and establishes an anchoring point for control linkage. There isn't any damper. The actuator will be anchored when the electromagnetic clutch is engaged (switched on).
Key Points to Check: appearance, installation, cable plug(s) and cable(s).
(2) Rudder Pedal Position Sensor GE-15 (29C): It receives the control position signals of the heading channel and transmits the signals to the computer.
Key Points to Check: appearance, blockage, cable plug(s) and cable(s).

Station 8

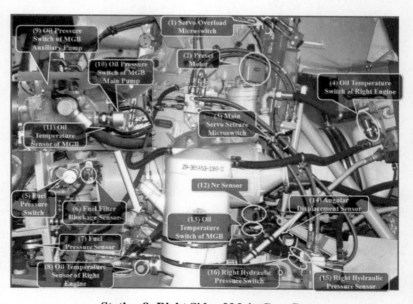

Station 8: Right Side of Main Gear Box

(1) Servo Overload Microswitch: It detects the maximum bearable load of the servo. Indicator "LIMIT" lights up when the load exceeds the limit. Press the button to test the circuit.
Key Points to Check: appearance, installation, cable plug(s) and cable(s).

(2) Preset Motor: It improves the response time of the engines to the change in collective pitch and adjusts the velocity in advance.
Key Points to Check: appearance, blockage, cable plug(s) and cable(s).

(3) Main Servo Seizure Microswitch(3D): It detects if the control and distribution valve of the main servo is seized. If so, it lights up indicator "SERVO" on 7α panel and indicator "ALARM" flashes. There is a test button on 6α panel to test if the system could work normally without hydraulic oil supply.
Key Points to Check: appearance, installation, cable plug(s) and cable(s).

(4) Oil Temperature Switch of Right Engine(40E): Its contactor is connected when the oil temperature reaches 125°C and indicator "OIL.TEMP" lights up.
Key Points to Check: appearance, installation, cable plug(s), cable(s) and leakage.

(5) Fuel Pressure Switch (14Q): It senses the fuel pressure at the outlet of the low-pressure pump. Indicators "FUEL" on 7α panel and "FUEL.P" on 3Q panel light up when the fuel pressure is lower than 0.2bar,
Key Points to Check: appearance, installation, cable plug(s), cable(s) and leakage.

(6) Fuel Filter Blockage Sensor: Indicators "FILT" on 3Q panel and "FUEL" on 7α panel light up when the pressure difference is over 140±20mbar.
Key Points to Check: appearance, installation, cable plug(s), cable(s) and leakage.

(7) Fuel Pressure Sensor: Its resistance changes with fuel pressure and generates corresponding voltage signals. The signals will be processed by the EPU and further displayed on the EPD (202E). Its range of measurement is 0-0.2MPa.
Key Points to Check: appearance, installation, cable plug(s), cable(s) and leakage.

(8) Oil Temperature Sensor of Right Engine (22E): Connected to the EPU, it works based on the change of resistance coefficient of nickel wire with the temperature.
Key Points to Check: appearance, installation, cable plug(s), cable(s) and leakage.

(9)(10) Oil Pressure Switch of MGB Main/Auxiliary Pump YKJ-4004-0.8 (65E, 66E): The oil pressure switch of MGB main pump is connected to indicator "MAIN.PUMP" while the oil pressure switch of MGB auxiliary pump is connected to indicator "AUX.PMP". They light up when the oil pressure is below 0.6bar.
Key Points to Check: appearance, installation, cable plug(s), cable(s) and leakage.

(11) MGB Oil Temperature Sensor (13E): It measures the oil temperature where it is mounted according to the change of resistance coefficient of nickel wire with the temperature. It is connected to the EPU.
Key Points to Check: appearance, blockage, cable plug(s), cable(s) and leakage.

(12) Nr Sensor (15E): It provides frequency signals of AC voltage in direct proportion to the rotation rate of phonic wheels. The signals will be transmitted to the EPD.
Key Points to Check: appearance, installation, cable plug(s), cable(s) and leakage.

(13) MGB Oil Temperature Switch (14E): The contactor of the switch will be connected when the oil temperature is up to $132.5℃±2.5℃$. In the meanwhile, indicator "OIL.TEMP" on 7α panel lights up.
Key Points to Check: appearance, installation, cable plug(s), cable(s) and leakage.

(14) Angular Displacement Sensor: It senses the position signals of the right front servo. It is exclusively used by the flight data system.
Key Points to Check: appearance, installation, cable plug(s) and cable(s).

(15) Right Hydraulic Pressure Sensor CY-YZ-0101-10 (4D): It transmits signals that indicate the pressure of the main hydraulic system (0-10MPa).
Key Points to Check: appearance, installation, cable plug(s), cable(s) and leakage.

(16) Right Hydraulic Pressure Switch CY-YCG-4 (6D): It senses hydraulic pressure and activates the circuit when the pressure reaches the max/min warning values. The indicator goes out when the hydraulic pressure rises to 2.5±0.5MPa and lights up when thehydraulic pressure drops below 1+0.5MPa.
Key Points to Check: appearance, installation, cable plug(s), cable(s) and leakage.

Station 9

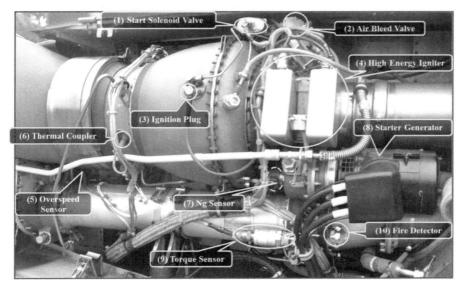

Station 9: Right Side of Engine (Right Engine Bay)

(1) Start Solenoid Valve RDK-16: It is a fuel injector solenoid valve that provides fuel to the combustion chamber at the initial phase of the start process. When the fuel pressure reaches 0.24MPa (2.4bar), it begins to inject air instead of fuel.
Key Points to Check: appearance, installation, cable plug(s), cable(s) and leakage.

(2) Air Bleed Valve ZQK-2: It is opened (air-operated) during engine start process (P_2/P_0 is below a certain value) to adjust air inflow and avoid surges. It is closed when Ng is between 90% and 100% (P_2/P_0 is over a certain value).
Key Points to Check: appearance, installation, cable plug(s), cable(s) and leakage.

(3) Ignition Plug BDH-15: It ignites the mixture of air and fuel in the initial phase of engine start.
Key Points to Check: appearance, installation, cable plug(s) and cable(s).

(4) High-energy Igniter DHZ-20B: The capacitive igniter converts low-voltage direct currents into high-voltage pulse currents (2000V) to generate electric sparks.
Key Points to Check: appearance, installation, cable plug(s) and cable(s).

(5) Overspeed Sensor GZ-18: During engine start, indicator "O/SPEED" goes out when Nf is up to about 25% and the monitoring system begins to work. When Nf is over 123.1%+2.5%, the sensor detects an overspeed and the overspeed cut-off solenoid valve shuts the engine down. A normal engine shutdown causes indicator "O/SPEED" to light up again. If the indicator goes out, it is required to press the reset button on the overspeed detection box to light up the indicator again.

Key Points to Check: appearance, installation, cable plug(s) and cable(s).

(6) Thermal Coupler GR-14: It detects the outlet temperature of the engines and transmits the data to the EPU for processing and display.

Key Points to Check: appearance, installation, cable plug(s) and cable(s).

(7) Ng Sensor GZ-17: It detects the rotation rate of the gas turbine and transmits the data to the EPU to monitor engine status.

Key Points to Check: appearance, installation, cable plug(s) and cable(s).

(8) Starter Generator GF-48: During engine start, it works as a motor to drive the engine. After engine start, it works as a generator to supply DC power to the helicopter.

Key Points to Check: appearance, installation, cable plug(s) and cable(s).

(9) Torque Sensor GY-20: It detects the output power of engines.

Key Points to Check: appearance, installation, cable plug(s), cable(s) and leakage.

(10) Fire Detector JUI-18: It detects fire of engines. The warning value is 200℃.

Key Points to Check: appearance, installation and cable(s).

Station 10

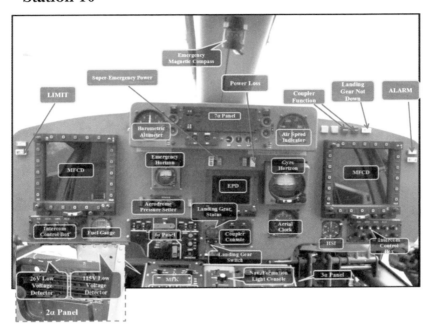

Station 10: Cockpit – Instrument Panel

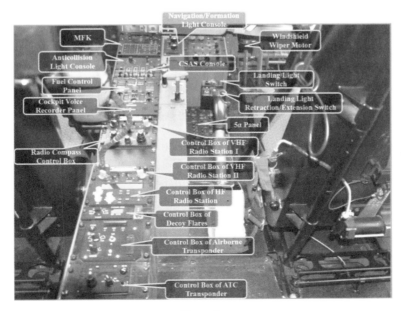

Console

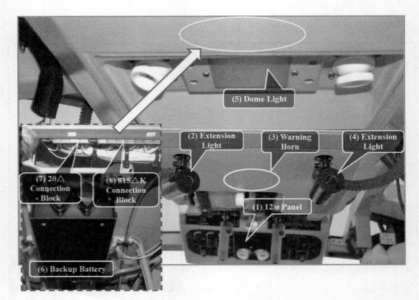

Cabin Dome

(1) 12α Panel: It is an electric control panel.

Key Points to Check: appearance and installation.

(2)(4) Extension Lights ZCD-16 (19L and 30L): They are used for map reading and partial illumination.

Key Points to Check: appearance, installation, cable plug(s) and cable(s).

(3) Warning Horn (20E): It generates two kinds of sound to warn the flight crew. High-pitch warning sounds indicate excessively high Nr while low-pitch warning sounds indicate low Nr. To trigger low-pitch sounds, Nr shall be between 170r/min and 345r/min; to trigger high-pitch sounds, Nr shall be over 380r/min.

Key Points to Check: appearance, installation, cable plug(s) and cable(s).

(5) Dome Light DD-7A (22L): It is for cockpit illumination.

Key Points to Check: appearance, installation, cable plug(s) and cable(s).

(6) Backup Battery CH-130 (124L): It is the emergency power source of the dome light and instruction light in a helicopter accident.

Key Points to Check: appearance, installation, cable plug(s) and cable(s).

(7) 20Δ Connection Block, (8) 815ΔK Connection Block: They are the connection blocks of the communication system, illumination system, oil system of MGB and engine monitoring and start systems.

Key Points to Check: appearance, installation and cable(s).

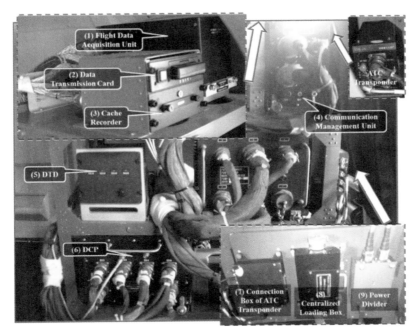

Equipment Rack in Cockpit

(1) Flight Data Acquisition Unit: It acquires and processes the data of the flight data system.

Key Points to Check: appearance, installation, cable plug(s) and cable(s).

(2) Data Transmission Card: It stores tactical data and other parameters to be bound into the system. It is capable of secret key erasing in an emergency.

Key Points to Check: appearance, installation, cable plug(s) and cable(s).

(3) Cache Recorder: It could record part of flight data with the flight data recorder and can be used to unload data quickly.

Key Points to Check: appearance, installation, cable plug(s) and cable(s).

(4) Communication Management Unit CH-130 (124L): CMU in short, it is pivotal in information processing as well as data link control and management of airborne platforms. It works with the avionic system, ultra-shortwave and short-wave radio stations to enable the data link system of airborne platforms.

Key Points to Check: appearance, installation, cable plug(s) and cable(s).

(5) DTD: Data transmission device is used to bind tactical data and other parameters (such as waypoints, route plans, types and quantity of external weapons and take-off weight).

Key Points to Check: appearance, installation, cable plug(s) and cable(s).

(6) DCP XKG-9B (7U): DCP (Display Control Processor) is the core unit of the IDCS. It receives data or signals output by all related mission devices for integrated processing so that the MFCDs (multi-functional color display) could display related pages of flight parameters and attack.

Key Points to Check: appearance, installation, cable plug(s) and cable(s).

(7) Connection Box of ATC Transponder: It is the connection box of the ATC transponder.

Key Points to Check: appearance, installation, cable plug(s) and cable(s).

(8) Centralized Loading Box (101R): It is used by the CMU to record data and videos. It has 4 serial ports and 2 USB ports.

Key Points to Check: appearance, installation, cable plug(s) and cable(s).

(9) Power Divider: It distributes signals from the transceiver of the transponder to front and rear antennas.

Key Points to Check: appearance, installation, cable plug(s) and cable(s).

Station 11

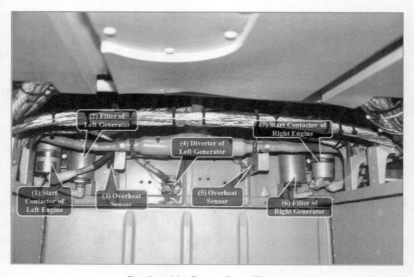

Station 11: Cargo Bay (Front)

(1)(7) Start Contactors of Left and Right Engines (25P and 24P): They connect the start circuits of starter generators during engine start.

Key Points to Check: appearance, installation, cable plug(s) and cable(s).

(2)(6) Filters of Left and Right Generators (27P and 26P): They filter AC interference during power generation and restrain radio jamming through wave filtering during engine start.

Key Points to Check: appearance, installation, cable plug(s) and cable(s).

(3)(5) Overheat Sensors (33W and 32W): The bimetallic strips of overheat detectors are disconnected when the temperature in the cargo bay reaches 110℃ and indicator "O/HEAT" on 7α panel lights up.

Key Points to Check: appearance, installation, cable plug(s) and cable(s).

(4) Diverter of Left Generator (33P): It measures the value of output current of the generator. The rated current value is 160A.

Key Points to Check: appearance, installation, cable plug(s) and cable(s).

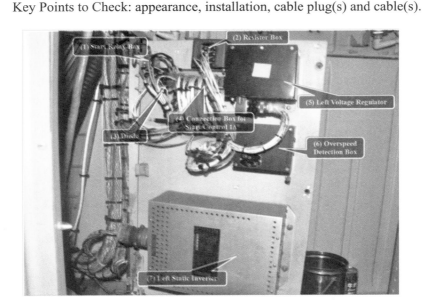

Left Electric Equipment Panel 21α

(1) Start Relay Box (7K): It is the relay box for engine start control.

Key Points to Check: appearance, installation, cable plug(s) and cable(s).

(2) Resistor Box: It is the resistor box (including 17W, 19W, 29W and 31W) for circuit compensation of the fire extinguisher system.

Key Points to Check: appearance, installation and cable(s).

(3) Diode (117P): It is the diode of the left voltage regulator.

Key Points to Check: appearance, installation and cable(s).

(4) Connection Box for Start Control 1Δ: It is the connection block of control circuits of the left engine and left power system.

Key Points to Check: appearance, installation and cable(s).

(5) Left Voltage Regulator TKB-3 (29P): It monitors the working status of the generator and regulates its output voltage. It has three functions: regulation (regulation of voltage during power generation and regulation of currents during engine start), control and protection.

Key Points to Check: appearance, installation, cable plug(s) and cable(s).

(6) Overspeed Detection Box (6K): It works together with the overspeed sensor to detect an overspeed. Normally, indicator "O/SPEED" lights up when Ng is below 25%. When an overspeed occurs (Ng>123.1%), it activates the overspeed cut-off solenoid valve to shut down the engine. There are also a test button and a reset button on the box.

Key Points to Check: appearance, installation, cable plug(s) and cable(s).

(7) Left Static Inverter PC-1000 (41X): It converts DC power into AC power for AC power consuming devices on board. It converts direct currents at 28V into two sets of alternate currents: 115V/400Hz and 26V/400Hz.

Key Points to Check: appearance, installation, cable plug(s) and cable(s).

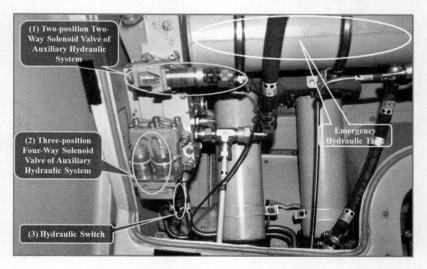

Auxiliary Hydraulic Regulator

(1) Two-position Two-way Solenoid Valve of Auxiliary Hydraulic System YDK-11 (15G): It is the control valve of the auxiliary hydraulic system. When it is switched on, the hydraulic oil in the pipes flows back to the tank. When it is switched off, the hydraulic oil could be pressurized.

Key Points to Check: appearance, installation, cable plug(s) and cable(s).

(2) Three-position Four-way Solenoid Valve of Auxiliary Hydraulic System YDK-10 (16G): It controls the retraction and extension of the landing gear by controlling the flowing direction of the hydraulic oil.

Key Points to Check: appearance, installation, cable plug(s) and cable(s).

(3) Hydraulic Switch YCG-4 (22G): It lights up indicator "AUX.HYD" when the auxiliary hydraulic pressure exceeds 2.5±0.5MPa.

Key Points to Check: appearance, installation, cable plug(s) and cable(s).

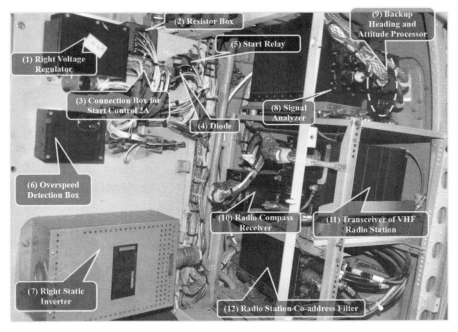

Right Electric Equipment Panel 22α

(1) Right Voltage Regulator TKB-3 (28P): It monitors the working status of the generator and regulates its output voltage. It has three functions: regulation (regulation of voltage during power generation and regulation of currents during engine start-up), control and protection.

Key Points to Check: appearance, installation, cable plug(s) and cable(s).

(2) Resistor Box: It is the resistor box (including 18W, 20W, 28W and 30W) for circuit compensation of the fire extinguisher system.

Key Points to Check: appearance, installation and cable(s).

(3) Connection Box for Start Control 2Δ: It is the connection block of the control circuits of the right engine and right power system.

Key Points to Check: appearance, installation and cable(s).

(4) Diode (118P): It is the diode of the right voltage regulator.

Key Points to Check: appearance, installation and cable(s).

(5) Start Relay Box (8K): It is the relay box for engine start control.

Key Points to Check: appearance, installation, cable plug(s) and cable(s).

(6) Overspeed Detection Box (5K): It works together with the overspeed sensor to detect an overspeed. Normally, indicator "O/SPEED" lights up when Ng is below 25%. When an overspeed occurs (Ng>123.1%), it activates the overspeed cut-off solenoid valve to shut down the engine. There are also a test button and a reset button on the box.

Key Points to Check: appearance, installation, cable plug(s) and cable(s).

(7) Right Static Inverter PC-1000 (40X): It converts DC power into AC power for AC power consuming devices on board. It converts direct currents at 28V into two sets of alternate currents: 115V/400Hz and 26V/400Hz.

Key Points to Check: appearance, installation, cable plug(s) and cable(s).

(8) Signal Analyzer (9T): It receives and processes the radar signals from the amplifier and detector, and further transmits the processed signals to IDCS.

Key Points to Check: appearance, installation, cable plug(s) and cable(s).

(9) Backup Heading and Attitude Processor (112F): It receives and processes magnetic heading and radio heading signals. Then it transmits processed signals to the IDCS to provide backup heading to the helicopter.

Key Points to Check: appearance, installation, cable plug(s) and cable(s).

(10) Radio Compass Receiver: It receives and processes signals from the combined antenna. Then it transmits the processed bearing signals to the HSI and audio signals to the intercom system. It could work in RECEIVE mode and Compass mode.

Key Points to Check: appearance, installation, cable plug(s) and cable(s).

(11) Transceiver of VHF Radio Station TKR128 (10R): Capable of signal reception and transmission, it is used for plain/secure communications between helicopters and between helicopters and the ground.

Key Points to Check: appearance, installation, cable plug(s) and cable(s).

(12) Radio Station Co-address Filter: It reduces the noises in spare bands during signal transmission of a radio station and restrains transmission/ reception interference to multiple co-address devices, so as to improve the ability of co-address communications for multiple communication devices.

Key Points to Check: appearance, installation, cable plug(s) and cable(s).

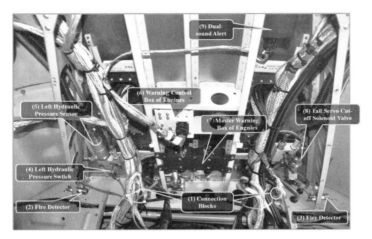

Ceiling of Cargo Bay (Forward)

(1) Ground Blocks (15N and 16N): They are the connection points of ground leads.
Key Points to Check: appearance, installation and cable(s).

(2)(3) Fire Detectors (25W and 26W): They detect fire in the cargo bay.
Indicator "CARGO.F" lights up when the temperature reaches 160℃ .
Key Points to Check: appearance, installation and cable(s).

(4) Left Hydraulic Pressure Switch YCG-4 (5D): It measures the hydraulic
pressure and activates the circuit when the pressure reaches the preset
max/min warning values. The indicator goes out when the hydraulic
pressure rises to 2.5±0.5MPa and lights up when the hydraulic pressure
drops below 1+0.5MPa.
Key Points to Check: appearance, installation, cable plug(s), cable(s) and leakage.

(5) Left Hydraulic Pressure Sensor CY-YZ-0101-10 (3D): It provides the
pressure signals of the main hydraulic system (0-10MPa).
Key Points to Check: appearance, installation, cable plug(s), cable(s) and leakage.

(6) Warning Control Box of Engines (815K): It is the power unit of audio
warning and supplies power to the audio warning device of the engine.
Key Points to Check: appearance, installation, cable plug(s) and cable(s).

(7) Master Warning Box of Engines (810K): It receives and processes signals
from the EPU and further transmits audio warning signals to headsets.
Key Points to Check: appearance, installation, cable plug(s) and cable(s).

(8) Tail Servo Cut-off Solenoid Valve YDK-12 (12D): It cuts off hydraulic oil
supply automatically when there is less than 2L of hydraulic oil in the tail
rotor hydraulic system (right hydraulic tank).

Key Points to Check: appearance, installation, cable plug(s) and cable(s).

(9) Dual-sound (gong sound) Alert BJQ-1 (813K): When Ng of one engine exceeds 100.4% of the maximum take-off power or the torque summation of two engines exceeds 102.5%, it transmits intermittent warning sounds to headsets. When one engine fails and the torque difference between two engines exceeds 25%, it transmits continuous warning sounds to headsets to warn the flight crew. Indicator "LIMIT" in the cockpit also lights up.

Key Points to Check: appearance, installation, cable plug(s) and cable(s).

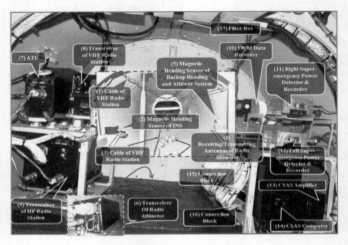

Cargo Bay (Backward)

(1)(3) Cables of VHF Radio Stations: They are connection cables.

Key Points to Check: appearance, installation, cable(s).

(2) Magnetic Heading Sensor of INS HJG-1A (42S): It measures the relative heading of the carrier to geomagnetic North Pole. It works with HJG-1A Inertial Navigation System to provide the relative angle of deflection to geomagnetic North Pole to strap-down inertial navigation system.

Key Points to Check: appearance, installation, cable plug(s) and cable(s).

(4) Receiving/Transmitting Antennas of Radio Altimeter (13S and 14S): They transmit and receive FM signals to measure altitude of the helicopter.

Key Points to Check: appearance, installation, cable plug(s) and cable(s).

(5) Magnetic Heading Sensor of Backup Heading and Attitude System (113F): It measures parameters of magnetic heading and provides backup heading.

Key Points to Check: appearance, installation, cable plug(s) and cable(s).

(6) Transceiver of Radio Altimeter GG0.6-1F (11S): It measures real-time absolute

altitude of the helicopter.

Key Points to Check: appearance, installation, cable plug(s) and cable(s).

(7) ATU (11R): It is the antenna tuning unit of the short-wave radio station. It matches the impedance between the transmitter and the antenna, so that the antenna could achieve maximum radiation power at any frequency.

Key Points to Check: appearance, installation, cable plug(s) and cable(s).

(8) Transceiver of VHF Radio Station (50R): As the transceiver of VHF FM, AM and anti-jamming radio stations, it can be used for air-to-air and air-to-ground voice communications.

Key Points to Check: appearance, installation, cable plug(s) and cable(s).

(9) Transceiver of HF Radio Station (10R): It receives and processes signals in the short wave band.

Key Points to Check: appearance, installation, cable plug(s) and cable(s).

(10) Flight Data Recorder: It is an anticollision recorder of flight data.

Key Points to Check: appearance, installation, cable plug(s) and cable(s).

(11)(12) Right/Left Super-emergency Power Detector and Recorder of (114E and 113E): When an engine holds super emergency power for more than 5 seconds, the warning flag on the super-emergency detector and recorder will turn from black to red. And the flash relay of this engine will be connected and the super-emergency power warning indicator lights up. There is also a life timer.

Key Points to Check: appearance, installation, cable plug(s) and cable(s).

(13) CSAS Amplifier FKJ-10 (19C): It controls pitch and roll double-motor actuators as well as the yaw actuator based on computer instructions. Also, it provides various signals needed by the computer and monitors the circuits.

Key Points to Check: appearance, installation, cable plug(s) and cable(s).

(14) CSAS Computer SJ-2C (16C): It processes signals from the gyro horizon, strap-down inertial navigation system and HADS. It also works with the coupler console and transmits processed signals to execution mechanisms to enable such functions as stability holding and auto-navigation of the helicopter.

Key Points to Check: appearance, installation, cable plug(s) and cable(s).

(15)(16) Connection Blocks (9ΔA1 and 9ΔB1): They are connection blocks.

Key Points to Check: appearance, installation and cable(s).

(17) Filter Box ELB-1 (650C): As a component of the CSAS, it is used to filter feedback position signals of the heading control actuator and HF interference of ±15V power.

Key Points to Check: appearance, installation, cable plug(s) and cable(s).

7 α Panel

TEST	GEN.1 — Left generator disconnected from bus bar (not working)	BUS.CPL — Bus bars coupled	GEN.2 — Right generator disconnected from bus bar (not working)	ROT.BK — Rotor braked	ENG.1 — Low oil pressure in left engine (<1.3bar) (Normal pressure: 0.8~3bar)	MGB.P — Low oil pressure in MGB(<1.3bar) (Normal pressure: 1.8~5bar)	ENG.2 — Low oil pressure in right engine (<1.3bar) (Normal pressure: 0.8~3bar)	CSAS / CSAS disengaged
	BAT.SW.1 — Battery disconnected from bus bar PP9	FUEL — Left/right fuel filter blocked or fuel pressure lower than 0.2bar	BAT.SW2 — Battery disconnected from bus bar PP8	DOORS — Cabin door(s) not closed	HYD.1 — Low pressure in right hydraulic system (<10bar) (Normal pressure: 60±8bar)	AUX.HYD — Overpressure in auxiliary hydraulic system (>30bar) (Normal pressure: 140bar)	HYD.2 — Low pressure in left hydraulic system (<10bar) (Normal pressure: 60±8bar)	C.P.L — Coupler fails
	INV.1 — Left AC power system fails	SHED.BUS — Bus bar shedding	INV.2 — Right AC power system fails	STEP — Step lowered down	OIL.TEMP — High oil temperature of MGB or Engines (MGB: >132.5°C, Engine: >125°C)	SERVO — Main servo seized	BAT.TEMP — High temperature of battery (>71±2°C)	
LAND. LT	GSS.HEAT — Heating system of velocity vector sensor fails (not working)	MAIN.PMP — Low pressure of main oil pump of MGB (<0.6bar)	PITOT — Heating system of pitot tube fails (not working)	O/HEAT — P$_2$ pipe of cargo bay overheated (>110°C)	CARGO.F — Fire in cargo bay (>160°C)	L/G.PUMP — Overpressure in emergency extension system of landing gear (pump in work) (>95bar/125bar)	FUEL.Q — Low fuel level (<18L)	GOV / Degrading failure of engines (minor failure of EECU)

Continued

DIM	EXT.L.H Left fire extinguisher empty	AUX.PMP Low pressure of auxiliary oil pump (<0.6bar)	EXT.R.H Right fire extinguisher empty	HORN Horn switched off (rotation speed of rotor, 170– 335, 380)	HYD.LEVEL Low hydraulic oil level of right hydraulic tank (<2L, cutoff)		TNG <u>Training with one engine inoperative (OEI)</u>
	CHIP.1 Metal chips of left engine detected	O/SPEED.1 Left engine overspeed detection system fails or Nf<25%	CHIP.2 Metal chips of right engine detected	O/SPEED.2 Right engine overspeed detection system fails or Nf<25%	DIFF.NG Power loss, OEI, <u>ΔNg>6%</u>	<u>GOV.2 Failure in right engine</u>	<u>RPM.365 Overspeed of rotor (>365r/ min)</u>

<u>GOV.1
Failure in left engine</u>

Note: The indicators underlined are indicators of Z-9WZ helicopters.

Positions of Devices in Group α

Code	Name	Position
1 α	Master Distribution Box	Radio Equipment Bay
2 α	Circuit Breaker Panel	Left of Control Console
3 α	Circuit Breaker Panel	Right Front of Control Console
4 α	Circuit Breaker Panel	Right Rear of Control Console
5 α	Circuit Breaker Panel	Right Central Part of Control Console
6 α	Power Control Panel	Left Bottom of Instrument Panel
7 α	Warning Indicator Panel	Upper Central Part of Instrument Panel
8 α	Instrument Panel	
9 α	Combined Instrument	MFCD(Z-9 Helicopter)
10 α	Voltage Selector	6 α
11 α	Master Warning Control Box	Radio Equipment Bay
12 α	Upper Electric Control Panel	Cockpit Ceiling
13 α	Cyclic Pitch Lever of Co-pilot	
14 α	Cyclic Pitch Lever of Commanding Pilot	
15 α	Collective Pitch Lever of Co-pilot	
16 α	Collective Pitch Lever of First Pilot	
17 α	Left(a) and Right (b) Master Warning Light	Instrument Panel
18 α	Left (a) and Right (b) Limit Indicator	Instrument Panel
19 α	Separation Connector	Upper Left of Cargo Bay
20 α	Separation Connector	Bulkhead X5630
21 α	Left Electric Equipment Panel	Cargo Bay
22 α	Right Electric Equipment Panel	Cargo Bay
23 α	Relay Box	Left Longeron of Bulkhead X1560
24 α	Relay Box	Right Longeron of Bulkhead X1560
30 α	Connection Box	Left of Bulkhead X1605
31 α	Connection Box	Below 23 α
35 α	Separation Connector	Lower Parting Surface of Vertical Stabilizer

Chapter 2   Power-on Check

Regulations on Power-on Check

1. Permission must be obtained from the flight engineer before ground power supply and power-on check;
2. Make sure that all power-distribution panels are well covered, all switches of power-driven devices are at proper positions and battery voltage is higher than 24 volts before switching on power;
3. The GPU (ground power unit) can only be connected to or disconnected from the helicopter when the main switches of the helicopter are at OFF positions. Do not disconnect the main switches without the permission of the person(s) who is (are) conducting power-on check on board.
4. Supply of ground power and power-on check are prohibited when:
   a. the helicopter is being refueled or defueled;
   b. the power system is malfunctioning or the parts and components of the power system are under check or being disassembled/reassembled;
   c. the helicopter or engine(s) is (are) being cleaned with gasoline;
   d. pipes and accessories of the fuel system are being assembled/disassembled or there is any leakage in the fuel system;
   e. voltage is not within normal range;
   f. there is strong thunder and lightning nearby;
   g. there is a possibility of short circuit;
   h. power-driven devices are being disassembled.

Preparations

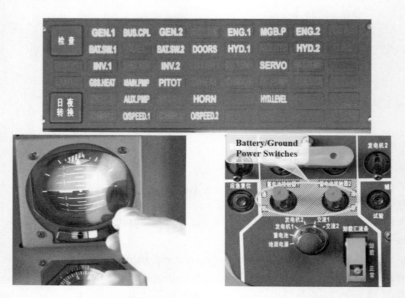

Pull out and hold the resetting knob of the gyro horizon (release the knob about 30 seconds later when the rotation speed of the gyro holds stable). In the meanwhile, switch on the two battery/ground power switches and press to extinguish indicator "ALARM". The indicators above shall light up on 7α panel.

Check on Voltage

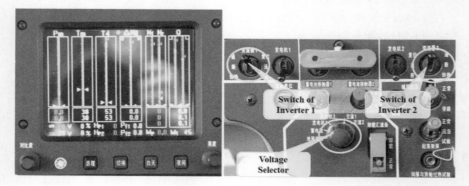

Engage inverters 1 and 2. When the EPD lights up, rotate the voltage selector to check if the voltages of the ground power unit, battery, inverters 1 and 2 are up to requirements.

Initiation of IDCS

(1) Put the main switch of the IDCS on MFK (multi-functional keyboard) to "STB" position. Next, set the switch to "ON" position when letters "MFK OK" stop flashing and letters "A IN" will be displayed in red on the MFCDs.

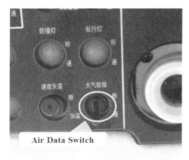

(2) Engage air data switch and letter "A" disappears.

Alignment of INS

Normal Alignment: Put INS switch to"NORM"position to activate normal alignment mode. In the first 4 minutes, the alignment mode is displayed as"FAST"; in the latter 4 minutes, the alignment mode is displayed as"NORM" instead. In 8 minutes, the display of"ALIGN IN PROGRESS"will change into"END OF ALIGN", calling an end to the alignment. Then put the"INS"switch to"NAV"position and the page of HSD will be displayed.

Fast Alignment: Put INS switch to"NAV" position and the alignment mode will be displayed as"FAST". About 3 minutes later, the HSD page will be displayed automatically, calling an end to the alignment.

Check on Barometric Altimeter and Fire Extinguisher System

FT/SET Knob

(1) Zero the needles of the barometric altimeter;

(2) Check the test switches of left and right fire extinguisher systems. Push the switch upward and the upper "FIRE" indicators light up; pull the switch downward and the lower "FAIL" indicators light up.

Check on 7α Panel

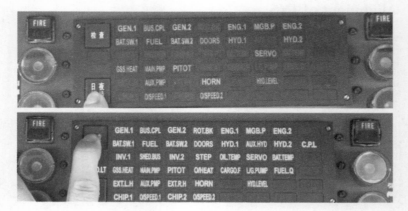

Press "TEST" button on 7α panel and 8 indicators will NOT light up (left and right "FIRE" and "FAIL" indicators, left and right "LIMIT" indicators and indicators of the transfer pump). Press "DIM" button and 7α panel will dim out. Press left and right "LIMIT" indicators and they will be brightened.

Check on MFCD

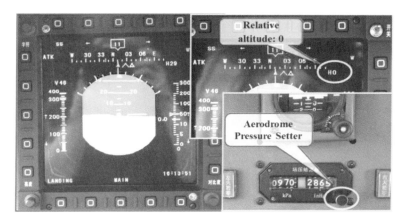

Adjust the aerodrome pressure setter until the relative altitude on VSD page is calibrated to zero and check if the aerodrome pressure indicated by the aerodrome pressure setter is the same as that indicated by the barometric altimeter.

Enter STATUS page: 1—IDCS, 3—HNS, 4—HADS, 6—EPU

Enter DATA page: 1—HNS, 2—HADS, 6—EPU, 9—TANK, 10—IDCS

Check if all the systems and devices above are working normally.

Check on Aerial Clock and EPD

Check on Aerial Clock: Check if the needles are set to zero, if there is any seizure in the movement of the needles and if the time is correct (press the button once to start time counting; press the button twice to stop time counting; press the button for the third time to reset time counting).

Check on EPD: Press "DAY" and "NIGHT" buttons and check system status.

Check on 6α Panel

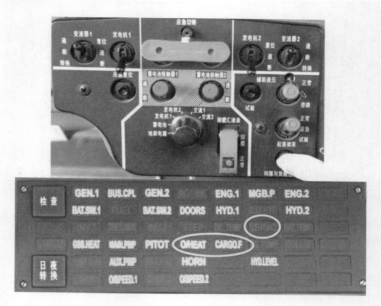

Press the test button of servo, cargo fire and overheat, and indicator "SERVO" goes out while indicators "O/HEAT" and "CARGO/F" light up.

Check on External Indicators

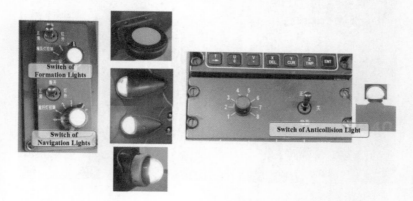

(1) Check on Anticollision Light: Switch on the anticollision light and it shall flash;

(2) Check on Navigation Lights: Switch on navigation lights and they shall light up (left one in red, right one in green and rear one in white);

(3) Check on Formation Lights: Switch on formation lights and they shall light up.

Check on Fuel Console

(1) Check if the pressure generated by booster pumps 1, 2, 3 and 4 is within the normal range (0.2−1.2bar); switch on booster pump 1 or 3 and the left "FUEL.P" indicator extinguishes; switch on booster pump 2 or 4 and the right "FUEL.P" indicator extinguishes; switch on left and right booster pumps at the same time and indicator "FUEL" on 7α panel extinguishes.

(2) Pull out and push the switch of the transfer pump leftward, and the left indicator for the transfer pump lights up, so does left "HI.LEV" indicator; pull out and push the switch of the transfer pump rightward, and the right indicator for the transfer pump lights up, so does the right "HI.LEV" indicator.

Check on 12α Panel

Engage the heating switch of the velocity vector sensor and indicator "GSS. HEAT" on 7α panel extinguishes; engage the heating switch of the pitot tube and indicator "PITOT" on 7α panel extinguishes (ground heating shall not exceed 2 minutes); engage the tail servo cutoff solenoid valve and indicator "HYD.LEVEL" on 7α panel extinguishes; engage cyclic trim switch and check if the trim release buttons (DBM) on cyclic pitch levers click clearly; engage horn warning switch and indicator "HORN" on the 7α panel extinguishes.

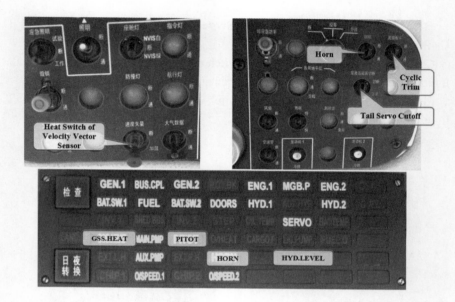

Check on Cyclic/Collective Pitch Levers and Rudder Pedals

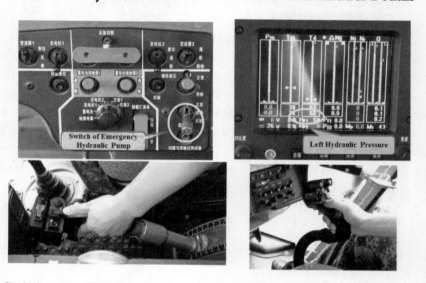

Switch on cyclic trim; set the switch of the emergency hydraulic pump to "TEST" position (to generate pressure in left hydraulic system); operate collective/cyclic pitch levers and rudder pedals to check if there is any retard or seizure. After the check, set the cyclic pitch levers and rudder pedals back to neutral positions and lower and lock the collective pitch lever.

Ground BIT of CSAS

Switch on the CSAS when:

(1) Failure warning flag of the gyro horizon retracts and three channel failure indicators (amber) of Lane 1 extinguish;

(2) Alignment of INS is completed and three channel failure indicators (amber) of Lane 2 extinguish;

(3) There is no light on the console;

(4) Inverters 1 and 2 and cyclic trim are switched on, with the switch of emergency hydraulic pump set to "TEST" position.

(1) Switch on Lane 1 and Lane 2; set BIT switch to "RUN" position to activate the built-in test of the CSAS.

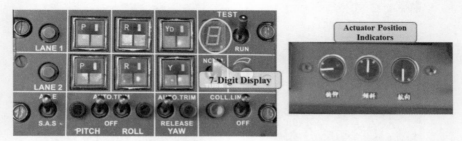

(2) Terminate the built-in test when "0" is displayed on the 7-digit display. When all the needles of the actuator position indicators go back to neutral positions, disconnect Lane 1 and Lane 2 simultaneously, and switch off the emergency hydraulic pump and cyclic trim.

Precautions in Ground BIT:

(1) Hydraulic oil must be supplied to the transmission system in ground BIT.

(2) BIT can't be interrupted.

(3) The CSAS can only be switched off when all the needles of the actuator position indicators go back to neutral positions.

(4) Switch off the CSAS first, then the hydraulic system.

(5) BIT is not allowed in the air.

Check on Coupler Console

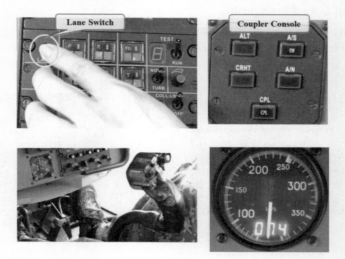

Switch on Lane 1 and Lane 2; switch on the coupler and then activate "A/S" on the coupler console. Now the helicopter stays in air speed holding state and cyclic pitch lever move forward automatically. When "74" is displayed on the air speed indicator, cyclic pitch lever stop moving.

Check on Illumination System

Set the emergency illumination switch to "NORM" position and main illumination switch to "ON" position; set the switch of cockpit illumination to "NVIS GREEN" and "NVIS WHITE" respectively; rotate the illuminating knobs for upper control panel and the instrument panel to the rightmost position. Check the illumination of the instrument panels, console, 4α, 5α and 12α panels and the extension lights. Switch on the landing light and indicator "LAND.LT" on 7α panel lights up, so does the landing light. Also, check the retraction and extension of the landing light (when there is a round-the-clock flight).

Check on Communication System

Put the intercom system in emergency mode and set the mode selector on the control box of TKR 128 radio station to "M/RE" (main receiver) position to activate the BIT. The test result will be displayed on the frequency/channel display. If the system works normally, "PASS" will be displayed, followed by the channel number.

Upon completion of the BIT, the radio station works in the displayed channel and noises or audio signals can be heard from the headsets. Set the working channel number based on the flight mission; set the frequency/mode selector at "READOUT" position; set the working frequency and press "SINGLE TONE" button to confirm. Then set the frequency/mode selector back to "PRESET" position and noises or audio signals can be heard from the headsets.

Set the squelch switch to "ON" position and do not operate other selectors. No noises shall be heard. If the system is receiving audio signals, no noises shall be heard during the intervals. Then set the squelch switch to "OFF" position.

Set the mode selector to "D/RE"(dual receivers) position, and the main receiver will be working in the preset mode while the survival receiver is working at the survival frequency and in the mode in correspondence. Noises or voices could be heard from the headsets.

Note: The survival receiver doesn't work in anti-jamming mode.

Set the control box in "M/RE" mode and set the frequency selector at "LIFE" position. Now the main receiver works at the survival frequency in correspondence with the preset frequency. The survival receiver does not work and noises can be heard from the headsets.

Set the control box in "M/RE" mode and set the frequency selector at "PRESET" position. Press "SINGLE TONE" button. The transmitter begins to work and single-tone signals can be heard from the headsets.

Note: Single tone function is not available in anti-jamming mode or when the frequency is between 30-88MHz.

Turn off the control box and the display window extinguishes. Set the frequency selector at "243" position and "243.000" shall be displayed. Now both the main receiver and transmitter are working at 243.000MHz in AM mode.

Set the frequency selector at "PRESET" position and set the control box in "M/RE" position. After the built-in test, press VHF button on the control box of JT-9 Intercom System while pressing the PTT button on the cyclic pitch lever and hold it at the second position. Send audio signals to the radio station working at the specified frequency and the speaker can hear his/her own voice from the headset. Release the PTT button and wait for the response.

After the check, reset the working frequency of the channel that has been operated and change the channel number back to that when the system was last shut down. Then turn off the radio station. Put the control box of the intercom system back to normal mode and reset other switches.

Check on Emergency Horizon

· Engage circuit breakers A4 on 4α panel and G2 on 5α panel. Check the working status of the emergency horizon. In normal temperature (5-20℃), the failure warning flag of the emergency horizon shall retract within 3 minutes;

· Pull out the resetting knob of the emergency horizon to zero it and the failure warning flag shall appear. Release the resetting knob and the failure warning flag shall disappear while the horizon remains zeroed;

· Disengage the circuit breakers and the failure warning flag re-appears.

Note: Check the following aspects 3 minutes after the gyro horizon is powered on:

· The artificial horizon shall be within the central circle of the aircraft symbol;

· Pitch angle indicator and bank angle indicator shall cross correctly within the central circle of the aircraft symbol;

· When the pitch angle indicator is parallel with the aircraft symbol, the depth of parallelism shall not exceed±3mm. The bank angle indicator shall not deviate from the zero scale mark by more than±3mm.

Chapter 3   Special Check

Check on KZW-3 Control Stability Augmentation System (CSAS)

1. Equipment Needed

One ground power unit (GPU) or an equivalent device

2. Preparations

- Open the covering caps of devices close to the CSAS.
- Plug in GPU.

3. Power-off Check

- Check every part of the CSAS in the following aspects:
  - ▷ general condition, deformation, cleanness, reliability and erosion (if any), labels and sealing (of sealed parts).
  - ▷ condition and connection of electrical connectors and the visibility of their labels.
  - ▷ mounting base of each part.
- Check the following accessories:
  - ▷ Circuit breakers: specifications and connections.
  - ▷ Actuator position indicators: electrical connection, needles and scales.
  - ▷ Warning indicators of the CSAS and the coupler: electrical connection and readability.
  - ▷ Coupler indicator: mechanical condition, markings and electrical connection.
  - ▷ Control buttons on cyclic and collective pitch levers: mechanical condition.
  - ▷ Air data calculater: nuts and pipes (general condition, cleanness, erosion

and fastening).

▷ Trim actuator: input linkage (general condition, cleanness, erosion, deformation, connection with the control linkage and rocking arms, conditions and cleanness of forks and cotter pins).

▷ Tandem actuator: connection between the shaft and the actuator on the crank, cotter pins, link rod (reliability, cleanness, erosion, general condition and fixation of nuts).

▷ Heading lever switch: mechanical condition and electrical connection.

▷ Conducting wires, line banks, earth link banks, earth points, electrical connectors, sheaths, shielding cases, connection between bulkheads and floor connectors, bonding jumpers and related mechanisms, wires in the sheaths, clamped sheaths and clamps connected to related mechanisms.

▷ General condition, erosion and reliability.

4. Power-on Check

Power on the helicopter and conduct the following check:

· Coupler Console of the CSAS:

▷ Condition of the mechanism, markings, scales and illuminating potentiometers.

▷ Illumination of indicators, markings, scales and if the indicators light up and go out correctly.

· Potentiometer:

▷ Condition of illuminating potentiometers of the instrument panel.

▷ If the potentiometer lights up and goes out correctly.

· Warning indicators of the CSAS: press "TEST" button on 7α panel to check if the warning indicators light up and go out correctly.

Note: Indicator "CSAS" on 7α panel is on when auto-trim switch on the CSAS console is being operated.

5. Finishing Touches

▷ Power off the helicopter and unplug the GPU.

▷ Close the covering caps.

BIT Procedures of KZW-3 Control Stability Augmentation System (CSAS)

1. Equipment Needed

A GPU or an equivalent device.

Note: A ground hydraulic power source or an equivalent device is needed when the CSAS is to work in only one mode for a long time.

2. Preparations

· Plug in the GPU to provide 28V DC power and 115V/400Hz and 26V/400Hz AC power to the helicopter.

· All actuators stay at neutral positions. The gyro horizon and inertial measuring unit of HNS work normally.

3. Test on Lane 1 and Lane 2

· Engage all channels and functions of the CSAS (all switches on the control box face forward);

· Put "TEST" switch at "RUN" position;

· During the whole test (about 30 seconds), the 7-digit display keeps flashing; 6 amber indicators (P1, P2, R1, R2, Y, YD) and 3 indicators for automatic trim ("AUTO TRIM") flash as the test goes on. After the indicators above flash for the second round, the code/digit on the display disappears, calling an end to built-in test.

· If the CSAS passes the test, "0" will be displayed at the end of the test.

· Any malfunction will be indicated by the corresponding code/digit. A maintainer could conduct trouble-shooting based on the digit and the maintenance manual.

· Put "TEST" switch back to "TEST" position.

· Indicators go out.

· Wait until the test ends.

4. Disconnect Lane 1

· Put "TEST" button at "RUN" position;

·During the test on the CSAS working in only one mode, indicators "TRIM" and "COLL LINK" shall light up at least once;

· After the test, put "TEST" button to "TEST" position. The needles of pitch actuator position indicator and roll actuator position indicator would jiggle as before but only with half the amplitude.

Note: The digit displayed is of no value for reference.

5. Disconnect Lane 2 and Reconnect Lane 1

· Disable "COLL LINK" and put "TEST" button at "RUN" position.

· Check the digit (1-7) displayed at the end of the test.

· Put "TEST" switch at "TEST" position and enable "COLL LINK".

· Reconnect Lane 2.

6. Flight Simulation

· Connect pin D of microswitch 9G with pin E of microswitch 10G.
· Put "TEST" switch to "RUN" position and BIT shall not start.
· Put "TEST" switch to "TEST" position.

7. Finishing Touches

· Remove the short-circuiting wire between 9G and 10G.
· Power off the helicopter and unplug the GPU.
· Decrease hydraulic pressure and then remove the ground hydraulic power source.

Test on AC Power System

1. Equipment Needed

None.

2. Preparations

· Turn on DC power system.
· Turn off inverters 1 and 2.
· Indicators "INV.1" and "INV.2" on 7α panel light up.

3. Testing Procedures

· Turn on inverters 1 and 2.
· Indicators "INV.1" and "INV.2" go out.
· Set voltage selector 10α at positions "INV.1" and "INV.2" to check if the voltage is 115V.
· Put switch "INV.1" at "OFF" position and then at "TRF" (transfer) position. 26V and 115V power systems will be restored and supply power.
· Put switch "INV.2" at "OFF" position and then at "TRF" (transfer) position. 26V and 115V power systems will be restored and supply power.

4. Finishing Touches

· Turn off inverters 1 and 2.
· Turn off DC power system.

Test on DC Power System

1. Equipment Needed

A 28V ground DC power source and a WQ1 0-40V adjustable voltage stabilizing DC power source.

2. Test on Generators

(1) Preparations.

· Plug in the GPU.

· Start generators.

· Put battery relays 49P (BAT.RLY1) and 50P (BAT.RLY2) at "ON" position, and indicators "BAT.SW1", "BAT.SW2", "GEN.1", "GEN.2" and "BUS.CPL" light up.

(2) Procedure.

· Unplug the GPU.

Note: The GPU can only be unplugged when the rotation speed of the generator reaches the rated value.

· Indicators "BAT.SW1", "BAT.SW2" and "BUS.CPL" go out.

Generator 1 (left):

· Put switch "GEN.1" at "RST" (reset) position and then at "ON" position. Indicator "GEN.1" goes out and indicator "BUS.CPL" lights up.

· Put voltage reading selector 10α at "GEN.1" position and the reading of voltage shall be 28.5V or 27.5V (it depends on the adjustment on the voltage regulator).

Note: If the reading of voltage is not correct, please adjust the voltage regulator.

· Put switch "BAT.RLY1" at "ON" position while switch "BAT.RLY2" remains at "OFF" position.

· Indicators "BAT.SW1" and "BUS.CPL" go out while indicator "BAT. SW2" keeps on.

· After 1s, put switch "BAT.RLY2" at "ON" position. Indicator "BAT. SW2" keeps on while indicator "BUS.CPL" remains off.

· Put switches "BAT.RLY1" and "BAT.RLY2" at "ON" position simultaneously and the corresponding indicators shall go out while indicator "BUS.CPL" lights up.

Generator 2 (right):

· Put switch "GEN.1" at "OFF" position and indicator "GEN.1" lights up.

· Put switch "GEN.2" at "RST" position and then at "ON" position. Indicator "GEN.2" goes out and indicator "BUS.CPL" lights up.

· Continue with the test on Generator 2 following the same testing procedures as those for Generator 1.

Test on Shedding System:

· Switch on two generators.

• Check generator voltage with a voltmeter (bus bars PP8 and PP9 are both live).

• Put bus bar shedding switch at "SHED" position and indicator "SHED" on 7α panel lights up. At the same time, power will no longer be supplied to bus bars PP10F (D10-D15) and PP13F (A6-A15) on 4α panel as well as bus bars PP15F (E1-E10) on 5α panel and PP12F (K1-K9) on 2α panel.

Note: Maintainers can also check if the bus bars are really shedded by checking if there is still power supplied to the devices connected to the circuit breakers on the panels.

Engine Shutdown:

• Put switches "BAT.RLY1" and "BAT.RLY2" at "OFF" position and all indicators go out.

• Put bus bar shedding switch at "NORM" position and close the safety cap.

3. Test on Battery and Ground Power Socket System

(1) Preparations.

• Make sure both engines are shut down.

• Put switches "BAT.RLY1", "BAT.RLY2", "GEN1" and "GEN2" at "OFF" positions.

• All indicators on 7α panel go out.

• Plug in the GPU.

Note: The GPU plug is always live whether it is plugged in or unplugged.

(2) Procedure.

Test on Ground Power System:

• Put switches "BAT.RLY1" and "BAT.RLY2" at "ON" position.

• Indicators "GEN.1", "GEN.2", "BUS.CPL", "BAT.SW1" and "BAT. SW2" light up.

• Put switch "BAT.RLY1" at "OFF" position while switch "BAT.RLY2" at "ON" position and do it the other way around.

• Indicators "GEN.1", "GEN.2", "BUS.CPL", "BAT.SW1" and "BAT. SW2" keep on.

• Put selector 10α at "EXT" (exterior power) position and check the voltage of GPU on the EPD.

• Put switches "BAT.RLY1"and "BAT.RLY2" at "OFF" position.

• Unplug the GPU.

Test on Battery System:

• Put switches "BAT.RLY1" and "BAT.RLY2" at "ON" position simultaneously.

· Put selector 10α at "BAT" (battery) position and check the voltage of the battery on the EPD.

· Indicators "BAT.SW1", "BAT.SW2" and "BUS.CPL" are off.

Note:

▷ If switches "BAT.RLY1"and "BAT.RLY1" are not engaged simultaneously, the battery relay switched on later will not be engaged because the short-circuit protection system will prevent the second battery relay from being engaged as a low voltage has been detected in the corresponding bus bar. Also, the bus bars coupling relay will not be engaged.

▷ During the whole test, battery temperature warning indicator shall not light up.

4. Test on Over-voltage Protection and Reset Circuitry

(1) Over-voltage protection.

· Start engines (it is advised to conduct the test together with the test on engine circuitry).

· As generators work with no load, press the small red button in the center of the radiator of the voltage regulator (28P or 29P) to create an over-voltage circuit. 1 second later, the output voltage of the generator rises to about 32.5V. Then the over-voltage protection circuit disconnects the excitation circuits of the voltage regulator and engine. Later the output voltage of the generator drops to below 1V and indicator "GEN.1" (or "GEN.2") lights up.

· Reset the voltage regulator with switch "GEN.1" or "GEN.2" on 6α panel. And the system voltage will be reset to 28.5V and the generator reverse current protector will be engaged.

(2) Emergency Reset.

· Put switches "BAT.RLY1" and "BAT.RLY2" at "OFF" position.

· Put switches "GEN.1" and "GEN.2" at "OFF" position, then at "ON" position.

· Press "EMERG.RESET" (emergency reset) button.

· The voltage of each generator displayed on an external voltmeter shall increase.

· Put switches "BAT.RLY1" and "BAT.RLY2" at "ON" positions.

· Indicators "GEN.1", "GEN.2", "BUS.CPL", "BAT.SW1" and "BAT.SW2" go out.

5. Test on Ground Power Socket System

· Unplug the battery.

· Connect a 0~40V adjustable DC power source to the binding posts of the helicopter ground power socket. The positive binding posts of the power source are connected to the positive and small binding posts of the ground power socket while the negative binding post of the power source is connected to the negative binding post of the ground power socket.

· Adjust the voltage of the adjustable power source to about DC 28V.

· Put switch "BAT.RLY1" at "ON" position.

· Engage circuit breaker 11P and supply power to the helicopter. An external voltmeter will indicate the voltage of the ground power source.

· Increase the voltage of the adjustable DC power source gradually.

· When the voltage reaches 33±1V, no power will be supplied to the helicopter and the screen of the EPD goes black.

· Adjust the voltage of the adjustable power source back to 28.5V. Still, no power will be supplied to the helicopter.

· Disconnect circuit breaker 11P.

· Reconnect circuit breaker 11P so that power can be resupplied to the helicopter and the EPD will display the voltage of the adjustable power source.

· Disconnect circuit breaker 11P to cut off power from the helicopter and the EPD goes black.

· Put switch "BAT.RLY1" at "OFF" position.

· Set the voltage of the adjustable power source to zero and disconnect it from the ground power socket.

· Reconnect the battery.

Adjustment and Balance Check on Generator Voltage of DC Power System

1. Equipment Needed

One C31-VA DC voltmeter at an accuracy of 1.5

Note: This procedure is conducted when it is the first ground operation adjustment or when a voltage regulator is newly installed.

2. Preparations

· Set the helicopter ready and make sure the battery is installed properly and fully charged.

· Open the cargo bay door.

· Open left and right electric equipment panels.

• Put switches "BAT.RLY1", "BAT.RLY2", "GEN1" and "GEN2" at "ON" position.

• Start engines and wait until their rotation speed reaches the rated value. Generators can only be connected with the power network after they enter the rated state (indicators "GEN1" and "GEN2" go out).

Note: When atmospheric temperature reaches or exceeds 35°C , it is advised to set the voltage of the helicopter to 27.5±0.5V. Otherwise, the voltage shall be set at 28.5±0.5V.

3. Testing Procedures (see the figure below)

(1) Voltage Check and Adjustment for No-load Generators.

• Put switches "GEN.1" and "GEN.2" at "OFF" position. Indicators "GEN.1" and "GEN.2" light up.

• Connect the voltmeter with left generator, then with right one: the positive wire is connected to the binding post (3) of the test socket of the corresponding voltage regulator while the negative wire is connected to the fuselage (fastening screw of the voltage regulator (2)).

• Record the voltage of each generator.

• If necessary, adjust the voltage to 28.5V, loosen the locking nut (4) and rotate the adjusting screw (1) to adjust the voltage (28.5±0.5V or 27.5±0.5V as required). Then fasten the locking nut and the voltage shall not change. If not, adjust again.

• Disconnect the voltmeter.

• Put switches "GEN.1" and "GEN.2" at "ON" position. Indicators "GEN.1" and "GEN.2" go out.

• Close left and right electric equipment panels.

• Close cargo bay door.

(2) Balance Check and Adjustment.

Put selector 10α on 6α panel at "GEN.1" position.

• Put switch "BAT.RLY1" at "OFF" position and indicator "BAT.SW1" lights up.

• Switch off Generator 1 and indicator "GEN.1" lights up.

• Check if the voltage of Generator 1 is the same as the adjusted voltage.

• Switch on Generator 1 and indicator "GEN.1" goes out.

• Switch off Generator 2 and indicator "GEN.2" lights up.

• Check the output current value of Generator 1.

• Switch on Generator 2 and indicator "GEN.2" goes out.

· Put switches "BAT.RLY1" and "BAT.RLY2" at "ON" position simultaneously. And indicators "BAT.SW1" and "BAT.SW2" go out.

Put selector 10α at "GEN.2" position.

· Put switch "BAT.RLY2" at "OFF" position and indicator "BAT.SW2" lights up.

· Switch off Generator 2 and indicator "GEN.2" lights up.

· Check if the voltage of Generator 2 displayed on the EPD is the same as the adjusted voltage.

· Switch on Generator 2 and indicator "GEN.2" goes out.

· Switch off Generator 1 and indicator "GEN.1" lights up.

· Check the output voltage of Generator 2. The value shall be the same as that of Generator 1.

· Switch on Generator 1 and indicator "GEN.1" goes out.

· Put switches "BAT.RLY1" and "BAT.RLY2" at "ON" positions simultaneously. And indicators "BAT.SW1" and "BAT.SW2" go out.

· Check the output voltage and current of Generator 2 and set selector 10α at "GEN.1" position.

· Check the output voltage and current of Generator 1.

· Normally, the voltage of Generator 1 shall be the same as that of Generator 2. The output current of each generator is about 40A (25%) and the output current difference shall not exceed 10A. Otherwise, re-adjust the voltage.

· Reduce the voltage of the generator under greater load to reduce its load.

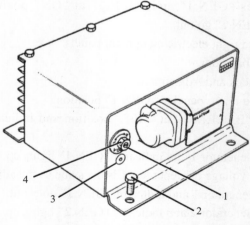

Voltage Regulator

Check on Battery Temperature Detector

1. Equipment Needed

A heating device (electrical oven is recommended), 2 containers (0.5L and 1L each), a thermometer (at the accuracy of 0.5℃), an avometer, conducting wires and connecting pins.

2. Preparations

· Remove the battery and put it on the working stand.

· Remove the battery cap.

· Remove the temperature detector (2 nuts and gaskets) from the cells.

· Add water to the container (0.5L) until the water fills 3/4 of its total volume. Then put the container on the oven.

3. Testing Procedures (see the figure below)

· Emerge the temperature detecting assembly in water and make sure the temperature detector (1) does not touch the container.

· Put the thermometer in water and draw the bulb (2) of the thermometer close to the temperature detector.

· Connect the avometer (3) or the continuity tester (4) with the two contactors of the temperature detector's socket. And the two contactors of the socket shall be connected with the two poles of the switch.

· Heat up the water gradually and keep the water temperature at 60℃ for 10 minutes to make sure the temperature detecting assembly acquires the same temperature as that of water.

· Continue to heat up the water and pay attention to the temperature reading when the temperature detector closes.

· As the temperature detector closes, the needle of the avometer sways from "∞" to "0" or the indicator of the continuity tester lights up. The closing point of a heated-up temperature detector is between 68℃ to 74℃ .

· Once the temperature detector closes, remove the oven and put the 0.5L container in the 1L one. Add cold water to the 1L container until the two containers acquire the same water level.

· Note down the temperature (cooling down) when the temperature detector is disconnected. When the temperature detector is disconnected, the needle sways back to "∞" or the indicator of the continuity tester goes out. The disconnecting point (temperature) of a cooling-down temperature detector is 4℃ -10℃ lower than the closing point (temperature) of a heated-up temperature detector.

· If the temperature detector is disconnected or closed at a temperature that is not within the limit of tolerance, repeat the procedures above.

4. Finishing Touches

· Take out the temperature detector from water and dry it up.

· Mount the temperature detector (2 nuts and gaskets) onto the cells of the battery.

· Put the battery cap back and install the battery.

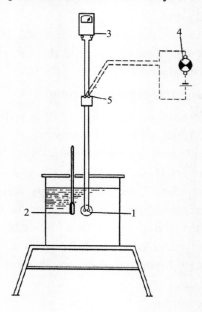

Check on Short-circuit Protection Plug-in Unit of DC Power System

1. Equipment Needed

None.

2. Procedure

Note: No device is allowed to be removed when engines are in operation on the ground.

(1) Right Power System.

· Connect both generators with the power network. Indicators "GEN.1", "GEN.2" and "BUS.CPL" will not light up.

· Disconnect circuit breaker C17 on 4α panel.

· Switch off Generator 2 and indicator "GEN.2" lights up.

· Reconnect circuit breaker C17.

· Put switch "GEN.2" at "RST" position.

· Check if the generator has been connected to the power network. Indicator "GEN.2" goes out.

(2) Left Power System.

· Conduct the same test as that on right power system. In this case, the circuit breaker to be operated is B17 on 4α panel.

Note: If the tests above are successfully completed, short-circuit protection plug-in units 40P and 41P can be seen as working normally.

Check on Diodes of Dual Power Supply Circuitry

1. Equipment Needed

28V DC ground power source.

2. The First Test

(1) Preparations.

· Disconnect the following circuit breakers on 4α panel: A4, A8, B1, B9, B14, B13, B16, B11, B12, A10 and A15.

· Disconnect the following circuit breakers on 5α panel: F1, F2, F3, F4, F5, F6, F7, F8, F9, and F10.

· Disconnect circuit breaker M2 on 2α panel.

· Start engines on the ground.

(2) Check and Trouble-shooting Measures.

· As Nr increases, check the warning sounds from the horn when Nr reaches the maximum and minimum values. If the sounds come back abnormal, change diode 28E on 2α panel or circuit breaker L3.

· Put inverter switches 38X and 39X at "ON" position and check if the 250VA AC power system outputs power and if indicators "INV.1" and "INV.2" go out. If not, check diode 46X and circuit breaker D8 on 4α panel.

· Disengage switch 8F on 12α panel and check the temperature of the pitot tube. Then engage switch 8F to see if the temperature rises. If not, check diode D6 on the 3α panel and circuit breaker D6 on 4α panel.

· Check the display on the fuel gauge 1Q. If there is no display or if the display is not correct, change the fuel gauge.

· Engage fuel transfer switch on 3Q panel and check if the fuel is transferred normally based on the reading on the fuel gauge. If not, check diode D4 on 2α panel and circuit breaker L6.

· Press the test button on 7α panel and check if the landing gear warning indicator "L.GEAR" 3G lights up. If not, check diode 56G on 2α panel and circuit breaker L9.

· Put switch 8E at "FIRE" position and check if fire warning indicator 2E ("FIRE") lights up. If not, check diode D3 on the 3α panel and circuit breaker L4 on 2α panel.

· Check if the CSAS works normally. If not, check circuit breaker D10 on 4α panel.

· Check if the emergency horizon works normally. If not, check circuit breaker G2 on 5α panel.

· Check if the super-emergency power system works normally. If not, check circuit breaker G3 on 5α panel.

· Check if the engine parameter acquisition and display system works normally. If not, check circuit breaker C5 on 4α panel.

· Check if HNS (Hybrid Navigation System) works normally. If not, check circuit breaker H6 on 5α panel.

· Press test button 2D. Check if indicators "CARGO.F" and "O/HEAT" on 7α panel light up. If not, check diode 3W on 32Δ connection box and circuit breaker C9 on 4α panel.

· Check if windshield wipers work normally. If not, check diode D1 on 3α panel and circuit breaker C16 on 4α panel.

· When power is supplied to the helicopter, check if the master warning indicator "ALARM" and most of the indicators on 7α panel light up. If not, check circuit breaker L8 on 2α panel.

· When power is supplied to the helicopter, engage the illumination switch on 12α panel and check if there is illumination in the instrument panel, instruments on the central console, circuit breaker panel, dome control panel and so on. If not, check circuit breaker D15 on 4α panel.

· When power is supplied to the helicopter, indicator "HYD.LEVEL" on 7α panel lights up. If not, check diode D28 and circuit breaker C11 on 4α panel.

· Disconnect the emergency illumination battery. Put the emergency illumination switch and cockpit illumination switch at "ON" position and check if the dome light lights up. If not, check circuit breaker C15 on 4α panel and diodes D1, D2, D3 and D4 on 126L support.

· Shut down engines on the ground and use GPU as the power source.

· Press test button 20G next to the GPU power socket and check if indicator 24G lights up. If not, unplug GPU and engage circuit breaker F8 on 5α panel. Then plug in GPU and press test button 20G. If indicator 24G lights up, replace diode D11 of 19G or 13P on 1α panel. If 24G still doesn't light up, unplug GPU, disconnect circuit breaker F8 and engage circuit breaker F9 on 5α panel. Then plug in GPU and press 20G. If 24G lights up, change diode D9 on 2α panel or circuit breaker G9 on 5α panel.

(3) Finishing Touches.

· Engage all the circuit breakers that have been disengaged.

· Put all the switches that have been operated in 2.2 back to their previous positions.

· Unplug GPU.

3. The Second Test

(1) Preparations.

· Disengage circuit breaker 13P in main power distribution box 1α.

· Disengage circuit breakers L1, L2, L3, L4, L6, L7, L8 and L9 on circuit breaker panel 2α.

· Disengage circuit breakers C9, C16, C14, D3, D4, D6, D8, D15, C11, C15, C5, and D10 on circuit breaker panel 4α.

· Disengage circuit breakers G9, G3, H6, and G2 on circuit breaker panel 5α.

· Start engines on the ground.

· Check if the CSAS works normally. If not, check circuit breaker A10 on 4α panel.

· Check if the emergency horizon works normally. If not, check circuit breaker A4 on 4α panel.

· Check if the super-emergency power system works normally. If not, check circuit breaker F6 on 5α panel.

· Check if the engine parameter acquisition and display system works normally. If not, check circuit breaker M2 on 2α panel.

· Check if HNS works normally. If not, check circuit breaker E9 on 5α panel.

(2) Check and Trouble-shooting Measures.

· As Nr increases, check the sounds from the horn when Nr reaches the maximum and minimum values. If the sounds come back abnormal, check diode 27E and circuit breaker B13 on 4α panel.

· Put switches 38X and 39X at "ON" position. Check the output voltage of

2×1000VA AC power system on the EPD and check if indicators "INV.1" and "INV.2" go out. If not, check circuit breaker A8 and diode 47X on 4α panel.

• Disengage switch 8F on 12α panel and check the temperature of the pitot tube. Then engage switch 8F and the temperature of the pitot tube shall increase. If not, check diode D9 on 3α panel and circuit breaker B1 on 4α panel.

• Check the display of the fuel gauge. If there is no display or if the display is not correct, replace the fuel gauge.

• Engage the fuel transfer switch on 3Q panel and check if the fuel is transferred normally. If not, check diode D1 on 2α panel and circuit breaker F3 on 5α panel.

• Press the test button on 7α panel and check if the landing gear warning indicator ("L.GEAR") lights up. If not, check circuit breaker F10 and diode 57G on 5α panel.

• Put switch 8E at "FIRE" position and check if warning indicator 2E ("FIRE") lights up. If not, check diode D2 on 2α panel and circuit breaker F5 on 5α panel.

• Put switch 9E at "FIRE" position and check if warning indicator 1E ("FIRE") lights up. If not, check diode D2 on 3α panel and circuit breaker F7 on 5α panel.

• Press the test button 2D and check if indicators "CARGO.F" and "O/HEAT" light up. If not, check diode 2W on 32Δb connection box and circuit breaker B9 on 4α panel.

• Check if windshield wipers work normally. If not, check diode D4 on 3α panel and circuit breaker B16 on 4α panel.

• Check if navigation lights light up normally. If not, check diode D7 on 4Δ connection box on 3α panel and circuit breaker B14 on 4α panel.

• When power is supplied to the helicopter, check if the master warning indicator "ALARM" and most of the indicators on 7α panel light up. If not, check circuit breaker B12 on 4α panel.

• When power is supplied to the helicopter, engage the illumination switch on 12α panel and check if there is illumination in the instrument panel, instruments on the central console, circuit breaker panel and dome control panel light up. If not, check circuit breaker A15 on 4α panel.

• When power is supplied to the helicopter, check if indicator "HYD. LEVEL" on 7α panel lights up. If not, check diode 29D and circuit breaker B11 on 4α panel.

• Connect the emergency battery for illumination. Put switches for emergency illumination and cockpit illumination at "ON" position and check if the dome light lights up. If not, check if the emergency battery for illumination works normally.

· Shut down engines on the ground.

· Press the test button 20G next to GPU power socket and check if indicator 24G lights up. If not, unplug GPU and engage circuit breaker G9 on 5α panel to draw power from the battery. Then plug in GPU and press 20G. If 24G lights up, replace diode D6 on 2α panel or circuit breaker F9 on 5α panel. If 24G still doesn't light up, unplug GPU, disengage circuit breaker G9 on 5α panel and connect the main power distribution box 13P to draw power from the battery. Then plug in the GPU and press 20G. If 24G lights up, change diode D10 in device 19G or circuit breaker F8 on 5α panel.

(3) Finishing Touches.

· Unplug GPU.

· Engage all the circuit breakers that have been disengaged.

· Put all the switches that have been operated back to their previous positions.

Check on Engine Fire Extinguisher System

1. Equipment Needed

An avometer.

2. Preparations

· Remove rear cowlings.

· Remove the cable plugs of the explosion tubes of two fire extinguishers.

· Remove the protection caps of the fire extinguisher system from the instrument panel.

· Switch on heliborne power system.

3. Check

At least 2 operators are needed in the following check (one in the cockpit and the other conducting the check).

Checking Procedures:

· Press the four buttons of the fire extinguishers in turn. After pressing each button, check voltages of binding posts A and C, B and C, or 1 and 3, 2 and 3 (markings vary based on the type of cable plugs) of the cable plug of the corresponding explosion tube:

▷ Press the first explosion button of left engine: the cable plug of yellow explosion tube of left fire extinguisher;

▷ Press the second explosion button of left engine: the cable plug of yellow explosion tube of right fire extinguisher;

▷ Press the first explosion button of right engine: the cable plug of gray explosion tube of right fire extinguisher;

▷ Press the second explosion button of right engine: the cable plug of gray explosion tube of left fire extinguisher.

· Check (general conditions, reliability, fastening, cracks, erosion and rust)

▷ cable plugs of explosion tubes

▷ cable plugs of pressure switches

▷ fire extinguishers (check the pressure with a pressure gauge based on the labels on extinguishers)

Atmospheric Temperature	−40	−30	−20	−10	0	+10	+20	+30	+40	+50
Pressure (bar)±10%	24	26	28.8	32.3	34.6	38	41.8	45.9	50.3	56.2
Pressure (MPa)±10%	2.4	2.6	2.88	3.23	3.46	3.8	4.18	4.59	5.03	5.62

▷ distribution pipes of fire extinguishers

▷ fire extinguisher nozzles on the rear fire proof wall

Check if there is any crack or fracture on the corners of the mounting bases of fire extinguisher bottles. If any, replace the mounting base(s).

4. Finishing Touches

· Switch off heliborne power system (generators and battery).

· Re-install the protection caps of fire extinguisher system onto the instrument panel.

· Remove the testing unit.

· Connect the cable plugs of the explosion tubes of two fire extinguishers.

· Re-install rear cowlings.

Check on By-pass Valve and Filter Blockage Indicator

1. Equipment Needed

A simulated filter core, a 28V ground DC power source, a fuel container (about 50L) and a fuel drain pipe.

2. Preparations

· Connect the ground power source to the helicopter.

· Open MGB and engine cowlings.

• Remove the locking wire and well place the fuel container. Disconnect the engine fuel-in pipe on the fire-proof baffle plate.

• Connect the fuel drain pipe.

3. Testing Procedures

Note: the test shall be conducted on the fuel-in pipe of each engine in turn.

• Remove the filter core.

• Replace the original filter core with the simulated one and put the simulated filter core into the fuel filter.

• Engage battery relays.

• Switch on booster pumps.

• Check the following points in turn to see:

   ▷ If filter blockage indicator "FILT" lights up.

   ▷ If the red ring on filter blockage indicator is fully visible.

Note: If indicator "FILT" doesn't light up while the red ring on filter blockage indicator is fully visible, press the indication cap and the indicator shall light up.

• Switch off booster pumps.

• Disengage battery relays.

• Remove the simulated filter assembly and re-install the original filter core.

• Reset the filter blockage indicator.

4. Finishing Touches

• Remove the fuel drain pipe.

• Re-install and fasten the fuel-in pipe.

• Remove the fuel container.

• Disconnect ground power source.

• Re-install MGB and engine cowlings.

Check on Low Fuel Level Switch of Feed Tank and Calibration of Fuel Gauge

1. Equipment Needed

A fuel drain pipe, a fuel tank with a fuel gauge with an accuracy within 0.5%, a graduated container (25L) and a fuel container (600L).

2. Preparations

• Lift up and level the helicopter.

· Open right front checking cap (to facilitate operation on the adjusting screw of fuel gauge amplifier) and open MGB and engine cowlings.

· Supply 28V DC power to helicopter.

· Place the fuel container and reduce the fuel in fuel tank groups to 48kg (about 60L).

· Replace the fuel-in pipe on the fire-proof baffle plate with a fuel drain pipe.

3. Testing Procedures (see the figure below)

Note: Low fuel level warning indicator may light up before the booster pumps are switched on but it has to go out after booster pumps enter into work.

· Drain the remaining 60 liters of fuel with heliborne pumps until low fuel level warning indicator lights up.

· Put a terminal of the fuel drain pipe into the graduated container and drain the fuel until the pumps begin no-load operation (dry run). Make sure there is more than 20 liters of fuel in the container.

· Rotate the corresponding adjusting screw on the fuel gauge amplifier to set the display of fuel gauge to 0kg.

· Add 400kg (about 506L) of fuel into the fuel tank groups.

· Rotate the corresponding adjusting screw (2) on the fuel gauge amplifier to set the display of fuel gauge to 400kg.

Note: The display of fuel gauge is supposed to be stable within 2 minutes after each adjustment on the adjusting screw.

· Remove the fuel drain pipes.

4. Finishing Touches

· Lower the helicopter down onto the ground.

· Connect and fasten the fuel-in pipe and remove the fuel container.

· Unplug GPU.

· Re-install MGB and engine cowlings and close the right front checking cap.

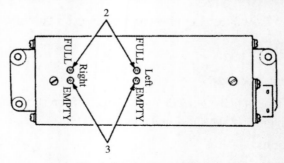

Fuel Gauge Amplifier

Check on Transfer Pump and High Fuel Level Switch

1. Equipment Needed

None.

2. Preparations

· Add 150kg of fuel into right fuel tank group.
· Fully fill left tank group with fuel.
· Supply 28V DC power to the helicopter.

3. Testing Procedures

· Check if relative indicators work normally.
· Pull and push the transfer pump switch rightward.
· Check:
  ▷ If rightward arrow indicator "→" lights up.
  ▷ If the displayed fuel amount of left fuel tank group decreases.
  ▷ If the displayed fuel amount of right fuel tank group increases.

Note: Indicator "HI.LEV" lights up when right fuel tank is full with fuel.

· Pull and push the transfer pump switch leftward.
· Check:
  ▷ If leftward arrow indicator "←" lights up.
  ▷ If the displayed fuel amount of right fuel tank group decreases.
  ▷ If the displayed fuel amount of left fuel tank group increases.

Note: Indicator "HI.LEV" lights up when left fuel tank is full with fuel.

· Stop fuel transfer.

4. Finishing Touches

· Disconnect power supply to the helicopter.

Check on Hydraulic System

1. Equipment Needed

None.

2. System Testing Procedures

· Switch on power system: indicators "SERVO", "HYD1" and "HYD2" light up.
· Press test button for servo and cargo fire "TEST SERVO+CARGO" and indicator "SERVO" goes out.

3. As rotors are rotating

· Indicators "SERVO", "HYD1" and "HYD2" go out.

· Press the test button on 7α panel. Indicators "SERVO", "HYD1" and "HYD2" light up. And the master warning light "ALARM" lights up.

· Check if the pressure of left and right hydraulic system is 6.0±0.2MPa (60±2bar).

Press the test button for servo and cargo fire "TEST SERVO+CARGO". Indicators "SERVO", "CARGO.F" and "O/HEAT" light up.

4. Test on By-pass Switch

· Press test button 28G on 6α panel and indicator "HYD.AUX" on 7α panel lights up.

· Press the test button and put switch 17G at "BY-PASS" position and indicator "HYD.AUX" goes out.

· Put switch 17G at "NORM" position.

Check on Indicators

1. Equipment Needed

None.

2. Preparations

· Put battery relays "BAT.RY1" and "BAT.RY2" at "OFF" position.

· All indicators on 7α panel, 3Q panel and 2G panel go out, so do indicators 17α and 3G.

· Press buttons "TEST" and "DIM", the indicators remain off.

· Button "DIM" is not live.

3. Testing Procedures

· Put battery relays "BAT.RLY1" and "BAT.RLY2" at "ON" position.

· Most indicators light up. Press buttons "TEST" and "DIM", and all indicators light up and the master warning light 17α "ALARM" lights up.

· Press indicator 17α "ALARM" and it goes out.

· Press and hold button "TEST" on 7α panel. All indicators on 7α panel, 3Q panel and 2G panel shall light up, so do indicators 17α and 3G.

· Press button "DIM" on 7α panel. Indicators on 7α panel, 3Q panel and 2G panel dim out while the brightness of indicators 17α and 3G remains the same.

· Release button "DIM" to retrieve daytime illumination. Disconnect 17α indicator and release button "TEST".

· Put battery relays "BAT.RLY1" and "BAT.RLY2" at "OFF" position and check if all related systems are reset to preparatory status.

Note: To avoid overheat of indicators, the indicator panel shall be dimmed during ground test or flight maintenance.

Check on Microswitches of Shock-absorbing Legs of Main Landing Gear

1. Equipment Needed

An avometer, insulating sealant and insulating varnish.

2. Preparations

· Jack the helicopter.

· Remove the door of landing gear bay (if any).

3. Adjustment on Microswitches

· Check if the shock-absorbing leg is fully extended.

· Loosen and remove 3 bolts (2) from the mounting base (1) to facilitate removal of the cover (4).

· Disconnect the cable plug of the microswitch.

· Connect the avometer to the socket (binding posts A and B).

· Loosen the locking nut (5).

· Rotate the screw (6) to identify the disconnecting points A and C of the microswitch.

· Rotate the screw (6) for 1 or 1.5 round(s) to leave proper clearance.

· Hold the screw (6) and fasten the locking nut (5).

· Drip a drop of varnish on the locking nut.

· Well position the cover (4) on the mounting base (1); fix the cover with 3 bolts (2) and 3 gaskets (3) whose threads have been applied with insulating sealant; lock the bolts and nuts of the microswitch with locking wires.

· Disconnect the avometer from the socket and reconnect the cable plug of the microswitch with the mechanism.

4. Finishing Touches

· Re-install the door of the landing gear bay (if any).

· Lower the helicopter down onto the ground.

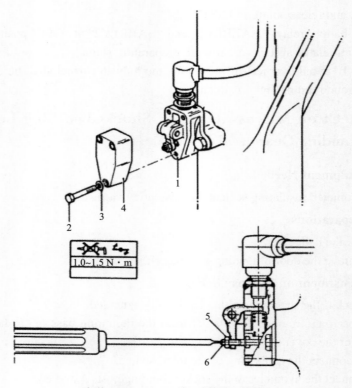

Adjustment on the Microswitch

Check on Retraction Interlock between Landing Gears

1. Equipment Needed

A main landing gear restraining panel, an avometer, a hydraulic jack and a 28V ground DC power source.

2. Preparations

a) Supply power to the helicopter.

b) Tilt the instrument panel backward.

c) Connect an avometer between binding post 1A of landing gear control switch 1G and the floor of the helicopter.

3. Testing Procedures

Note: The hydraulic system shall not run with excessive load during the test. Check if the landing gear control switch is secured by safety pins.

a) Install the main landing gear restraining panel on left main landing gear.

b) Jack the helicopter.

c) Check:

　· If the shock-absorbing legs of nose landing gear and right main landing gear are extended.

　· If nose landing gear is centered and locked.

d) Check if the reading of the avometer is zero.

e) Lower down the helicopter.

f) Remove the main landing gear restraining panel from left main landing gear and install the panel on right main landing gear.

g) Jack the helicopter.

h) Check:

　· If the shock-absorbing legs of nose landing gear and left main landing gear are extended.

　· If nose landing gear is centered and locked.

i) Check if the reading of the avometer is zero.

j) Lower down the helicopter.

k) Remove the main landing gear restraining panel from right main landing gear.

l) Jack the helicopter.

m) Check if all shock-absorbing legs are extended.

n) Turn the wheels of nose landing gear so that they deviate from the symmetrical center line by more than 2 degrees.

o) Check if the reading of the avometer is zero.

p) Center and lock nose landing gear and put landing gear switch at "EMERG" position.

q) Check if the reading of the avometer is zero.

r) Center and lock nose landing gear and lower the helicopter down on the ground.

s) Make sure rocker arm (2) keeps clear of contactor (1) of the microswitch.

4. Calibrating Procedures

a) If rocker arm (2) touches contactor (1) of the microswitch, re-adjust the microswitch.

b) If the reading of the avometer is not as zero as stipulated in steps 3d), 3i) and 3o), re-adjust related microswitches.

c) Conduct the test again after the microswitch(es) is(are)re-adjusted.

d) If the reading of the avometer is still not as zero, replace the microswitch.

e) Conduct the test on newly-replaced microswitch(es).

f) If the reading of the avometer is not as zero as is stipulated in step 3q), replace the emergency switch of the landing gear.

5. Finishing Touches

a) Disconnect and remove the avometer.

b) Lower down and lock the instrument panel.

c) Unplug the GPU.

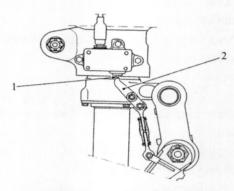

Microswitch of Landing Gear

Calibration of Compass of Back-up Heading and Attitude System

1. Equipment Needed

A 28V ground DC power source.

2. Preparations

Connect the 28V ground DC power source with the helicopter.

3. Testing Procedures

(1) Requirements.

• Compass calibration is required before the helicopter is delivered from the factory or before the first flight after a large-scale modification on the helicopter.

• Compass calibration is required upon replacement of processor of the back-up heading and attitude system.

• Compass calibration shall be conducted after the back-up heading and attitude system passes the power-on check.

• The compass shall be calibrated on plain sites that are 200 meters away from steel-made mechanisms, underground cables, large metallic pipes, high buildings, high-voltage cables and any physical entities prone to induce changes to geomagnetic field.

· Airborne devices (accompanying devices, products and accessories) shall come with complete supporting devices and shall be properly placed.

· All airborne electrical devices shall be working normally before the calibration.

· Power system shall not be switched off during the calibration, or it will generate great error in magnetic heading.

(2) Check and Calibration.

Preparations:

· Place the helicopter on the calibration site.

· Unlock the control system of the helicopter. Set all movable devices such as control mechanisms, rudder pedals, cyclic pitch levers and collective pitch levers at neutral positions (cruising flight positions).

· Switch on all airborne electrical devices and make sure they work normally.

Compass Calibration:

· Set the calibration switch on the processor of back-up heading and attitude system rightward to "CALI." position.

· Power on the helicopter and the needle of HSI (horizontal situation indicator) keeps turning clockwise nonstop, indicating that the system is being calibrated.

· When the system is being calibrated, push the helicopter for 2 rounds at a constant speed on the calibration site (720 degrees). Each round shall cost not less than 1 minute.

· Put the calibration switch on the processor of back-up heading and attitude system leftward to "NORM" position.

· Wait until the needle of HSI stops turning (heading is 0 degree), calling an end to the calibration. Then power off the helicopter.

4. Finishing Touches

Switch off all live devices and remove the GPU.

Calibration of HJG-1A Magnetic Heading Sensor of HNS

1. Equipment Needed

None.

2. Requirements

· Compass calibration is required before the helicopter is delivered from the factory or before the first flight after a large-scale modification on the helicopter.

· Compass calibration and zeroing is required upon replacement of the magnetic heading sensor. Zeroing is required upon replacement of the inertial measuring unit.

· The compass shall be calibrated on plain sites that are 200 meters away from steel-made mechanisms, underground cables, large metallic pipes, high buildings, high-voltage cables and any physical entities prone to induce changes to geomagnetic field.

· Airborne devices (accompanying devices, products and accessories) shall come with complete supporting devices and shall be properly placed.

· All airborne electrical devices shall be working normally before the calibration.

3. Preparations

· Place the helicopter on the calibration site.

· Switch on all airborne electrical devices and make sure they work normally.

4. Compass Calibration

(1) Compass Calibration.

· Press selector keys D3, R1 and L1 on MFCD to enter into the navigation page.

· Press "M-CAL" to enter the page for calibration of magnetic heading sensor (as in the figure below).

· Press "M-CAL" to start the calibration and "M-CAL: ON" would be displayed on the display.

· At the same time, push the helicopter for 1 round at a constant speed within 2 minutes.

· In 2 minutes, HNS finishes the calibration and "M-CAL: OFF" is displayed.

Note: The calibration result is classified into 9 levels. If the displayed result is "CAL LEVEL-X/Y: 9", HNS passes the calibration. If not, another calibration is needed.

(2) Warp Correction.

· After compass calibration, align the symmetrical axis of the helicopter with the 0° axis of the calibration site (magnetic heading: true north).

· Read the magnetic heading output by the magnetic heading sensor displayed on the compass calibration page for magnetic heading sensor.

· Calculate the value of warp.

· Press key selector "WARP" and input the warp correction value of magnetic heading sensor with MFK (multi-functional keyboard). For example, if the magnetic heading value of magnetic heading sensor (M-HDG) is 2 degrees, input 2; if M-HDG is 358 degrees, input -2 degrees.

· Press "ENT" to confirm and call an end to the zeroing of magnetic heading sensor.

5. Finishing Touches

· Put switches "INS", "IDCS" and "HNS" at "OFF" position.

· Switch off all airborne live devices.

· Unplug the GPU.

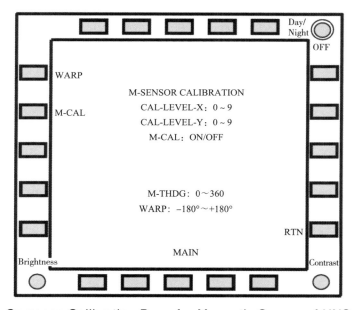

Compass Calibration Page for Magnetic Sensor of HNS

M-HDG:Magnetic Heading

WARP:Zero Offset Correction Value of Magnetic Sensor

CAL-LEVEL-X/Y:X/Y Axis Correction Level

M-CAL:Magnetic Calibration Enabling

Note:The calibration ends automatically 2 minutes after being enabled. Press "M-CAL"
      again to stop the ongoing calibration.

Check on Engine Super-Emergency Power Control System

1. Equipment Needed

A 28V ground DC power source.

2. Preparations

· Note down the working time recorded by the engine power detector and recorder and plug in the GPU.

· Put battery relays 49P and 50P at "ON" positions. Indicators "BAT. RLY1", "BAT.RLY2", "GEN1", "GEN2" and "BUS CPL" light up.

3. Testing Procedures

· Engage switch 37E for horn warning and fully turn up the volume of the compass with the knob on the intercom control box.

· Start up one engine according to stipulated procedures and release the start button when Ng reaches 40%. Push the throttling lever to flight position. When torque difference slightly exceeds 25%, two super-emergency power warning indicators light up and the continuous sounds of a gong ring in the headset. Then put the throttling lever back to idling position to decrease torque difference to less than 25%. Two super-emergency power warning indicators go out and the warning sounds in the headset stop.

· Start up the other engine and push the throttling lever to flight position to increase torque difference to over 25%. Two super-emergency power warring indicators light up and continuous sounds of a gong ring in the headset. Then pull back the throttling lever to shut down the engine.

· Disconnect engine super-emergency control switch 121E on 12α panel. Push the throttling lever to flight position again to increase torque difference to over 25%. The super-emergency power warning indicators will not light up and continuous sounds of a gong will not ring up in the headset.

· Pull back the throttling lever to shut down the engine. Open MGB cowlings and check the working time recorded by the engine power detector and recorder. The time recorded shall be the same as the actual working time (timing starts when Ng is over 50%) and the warning flag will not turn.

4. Finishing Touches

· Unplug the GPU.

· Close MGB cowlings.

· Engage engine super-emergency control switch 121E on 12α panel and finish with a lead sealing with a locking wire.

Note: The test shall be conducted when the helicopter is moored on the test stand.